Student Study

to accompany

Principles of Genetics

Robert H. Tamarin
University of Massachusetts–Lowell

Prepared by
Deborah C. Clark
Middle Tennessee State University

WCB McGraw-Hill

Boston Burr Ridge, IL Dubuque, IA Madison, WI New York San Francisco St. Louis
Bangkok Bogotá Caracas Lisbon London Madrid
Mexico City Milan New Delhi Seoul Singapore Sydney Taipei Toronto

WCB/McGraw-Hill

*A Division of The **McGraw-Hill** Companies*

Student Study Guide to accompany
PRINCIPLES OF GENETICS, SIXTH EDITION

1 2 3 4 5 6 7 8 9 0 BKM/BKM 9 0 9 8

ISBN 0-697-35464-4

www.mhhe.com

ACKNOWLEDGMENTS

I am indebted to a number of individuals for their help and encouragement during the preparation of this study guide. Dr. Philip Mathis of Middle Tennessee State University generously shared his support and expertise by reviewing and critiquing the manuscript. The undergraduates at Middle Tennessee State University, especially Shuhanna Carter, Jennifer Raymond, and Michael May, provided valuable support, proofreading, critiquing, and field-testing of the manuscript. The format of the study guide was selected because it best met the needs of the undergraduate genetics student. This guide is written, after all, for the student. Thank you is also extended to Mike Prater and Kathy Naylor at McGraw-Hill. Finally, I thank my spouse John Clark for his support and encouragement while I was writing and teaching. Any misprints, mistakes, incorrect calculations, or misleading information are solely my responsibility.

D. C. C.

PREFACE

I can still remember the first time I learned about a Monk named Gregor Mendel. In high school, I was fascinated by the simple rules that defined a field of study known as genetics. I never would have guessed that I would later make a career teaching the subject that so fascinated me. My excitement about the study of genetics has only intensified since that time. As you work through your textbook and this study guide, you will come to understand why so many scientists choose to focus on genetic questions. Along the way you will sharpen your problem solving abilities and expand your knowledge about a relatively new field that is destined to be the science of the next century. Hopefully, by the end of this course, you will understand and share in the excitement I first felt when I learned how to predict the results of a cross using a Punnett square. I hope that this study guide will assist you during your investigations and will help to alleviate some of the frustration that always accompanies the process of learning a new way to view the natural world.

How to Use the Study Guide

Your genetics course will no doubt differ greatly from the other biology courses you have taken. Genetics is a problem-solving discipline. This study guide emphasizes genetics as an on-going process of inquiry and investigation and promotes the development of critical analysis skills. You cannot passively learn the subject matter presented in your textbook and study guide. You must actively participate in the learning process by working problems and analyzing genetic data. The old saying *practice makes perfect* definitely applies when taking a course in genetics. You should begin to study by reading the relevant chapters of your textbook. Ideally, you should read the chapter *before* the material is covered in lecture. After the lecture, read the *key genetic concepts* section of the study guide to review. Do not attempt to use this section in place of reading the textbook or attending lectures. Once you feel confident that you understand the concepts covered, review the *problem-solving hints* provided in the guide. Finally, begin working the problems in the textbook and in the guide. The *practice problems* and *general questions* found in the study guide are challenging: most will require you to apply general principles and to analyze genetic data. While detailed solutions are provided, don't be tempted to look at the answers until you have attempted to work the problem in its entirety. Remember that the answers will not be available on the examination! If you get stuck, don't get frustrated and don't give up! Try to identify where you are having problems and re-read the related section in the textbook. Re-examine the problem solving hints in the study guide. If necessary, read over the explanation of how to work the problem in the solutions section. Discuss any problems you have with your instructor. Prior to the examination, review the material in the study guide to refresh your memory. If you have followed the above suggestions, you should be ready for the examination.

How to "Survive Genetics"

A common question: *How do I survive genetics?* The answer: *you put in your best effort.* As you begin your studies, you may find the simple hints outlined below will help to point you in the right direction.

- **Read the assigned pages.** The textbook you are using, ***Principles of Genetics*** by R. Tamarin is clearly written and easy to understand. Review the textbook material before you go to class. The material will be easier to follow during the lecture if you are familiar with the subject matter to be covered. Read the study guide summary *after* the lecture and before an examination to review the main concepts. Remember that this guide is a *companion* to the textbook.
- **Attend all of the lectures.** Each instructor will emphasize different areas of genetics. You will want to be familiar with the course objectives and requirements so you can focus your study efforts. Your instructor may also explain concepts differently than the textbook or this study guide. Obtaining different perspectives on the same material will help you to assimilate the material and achieve a more complete understanding of the subject.
- **Ask questions.** Throughout your education you will be encouraged to ask questions. Begin asking them now! The best way to achieve a full understanding is to ask questions. You must remember that your instructor wants you to successfully complete the course. However, he/she cannot help you if they are unaware that you are having problems. Before going to talk with your instructor, identify the concepts or procedures you are having difficulty understanding or performing. In class, ask the instructor to clarify points if necessary. Don't be afraid to ask for help.
- **Learn the vocabulary.** Geneticists use a number of terms that will be new to you. You must learn this vocabulary to understand the concepts being discussed and to solve problems. Relevant terms appear in bold in the study guide's review of genetic concepts.
- **Work all of the problems.** As mentioned above, genetics involves problem solving. The study guide provides hints you will find useful when working the problems in the textbook and study guide. Re-work all assigned problems prior to the examination.
- **Form a study group.** Many students are hesitant about studying in a group but I highly recommend that you give it a try. If you are having problems, you may feel more comfortable working with a group of friends in an informal setting. If you are comfortable working problems, a deeper understanding can be gained by teaching your classmates how to approach the subject matter. You never truly learn a subject until you must teach it to others. Invite your instructor to join your meeting if she/he is available. Remember, don't be a passive participant!
- **Have fun.** Learning should be an enjoyable experience for you. Focus more on learning about a fascinating subject and less on the grade! If you enjoy learning, you will achieve more than an "A." A positive attitude will contribute to your success in genetics, in your education, and in your life.

THE PIONEERS OF GENETICS

The contributions of many scientists have created and advanced the field of genetics over the past 100 years. As you encounter each below listed individual during your studies, identify their major contribution to the current understanding of genetic principles.

G. Mendel	G.W. Beadle	E.L. Tatum	C. Bridges
M.F. Lyon	T.H. Morgan	A. Sturtevant	B. McClintock
H.B. Creighton	C. Stern	J. Lederberg	F. Griffith
E. Wollman	M. Delbrück	J.D. Watson	F.H.C. Crick
M. Wilkins	R. Franklin	O. Avery	A. Hershey
M. Chase	E. Chargaff	M. Meselson	F.W. Stahl
M.W. Nirenberg	S. Ochoa	A. Jeffreys	W. Gilbert
F. Sanger	W. Arber	H. Smith	D. Nathans
P. Berg	F. Jacob	F. Monod	S. Benzer
S. Luria	L. Margulis	J.F. Crow	R.A. Fisher
S. Wright	C. Darwin	S.J. Gould	R.C. Lewontin

CHAPTER 1
INTRODUCTION

Key Genetic Concepts

The Scientific Method: All scientists, including geneticists, use the scientific method as a tool to understand and investigate questions about the natural world. The essential elements of the scientific method are observation, hypothesis formulation, prediction, experimentation, and analysis.

Sub-Disciplines of the Field of Genetics: The field of genetics can be subdivided into three distinct areas: classical, molecular, and evolutionary. Each sub-discipline focuses on different types of questions and at different levels of analysis.

- **Classical Genetic Studies** focus on the transmission of traits from parents to offspring. This sub-discipline encompasses the study of Mendel's principles (Chapter 2); the cell cycle, mitosis and meiosis (Chapter 3); sex determination and linkage (Chapter 5); genetic mapping (Chapters 6 and 7); and cytogenetics (Chapter 8).
- **Molecular Genetic Studies** focus on the genetic material. This sub-discipline includes studies of DNA structure, chemistry, and replication (Chapters 9 and 14); gene expression (Chapters 10 and 11) and its control (Chapters 13 and 15); DNA cloning (Chapter 12); and extrachromosomal inheritance (Chapter 17).
- **Studies of Evolutionary Genetics** focus on the changes in gene frequencies observed in populations of organisms. This sub-discipline includes studies of quantitative traits and heritability (Chapter 18); Hardy-Weinberg equilibrium (Chapter 19); the forces leading to evolutionary change (Chapter 20); and the process of speciation (Chapter 21).

General Questions

Question 1-1. Classify each pioneering scientific investigation described below according to the area of study (classical, evolutionary, or molecular genetics).

a. Genetic control of biochemical reactions in *Neurospora* (G.W. Beadle and E.L. Tatum)

b. The genetics of sex in *Drosophila* (C. Bridges)

c. The chromosomes in heredity (W.S. Sutton)

d. Sex limited inheritance in man (T.H. Morgan)

e. Sex chromatin and gene action in the mammalian X chromosome (M.F. Lyon)

f. A correlation of cytological and genetical crossing over in *Zea mays* (H.B. Creighton and B. McClintock)

g. The effects of unequal crossing over at the *Bar* locus in *Drosophila* (A. Sturtevant)

h. Studies on the chemical nature of the substance inducing transformation in *Pneumococcal* types (O. Avery, C. MacLeod, and M. McCarty)

i. Molecular structure of nucleic acids: A structure for deoxyribonucleic acid (J.D. Watson and F.H.C. Crick)

j. Independent functions of viral protein and nucleic acid in growth of bacteriophage (A. Hershey and M. Chase)

k. Genetic regulatory mechanisms in the synthesis of proteins (F. Jacob and J. Monod)

l. Artificial transmutation of the gene (H.J. Muller)

m. The role of inheritance in behavior (R. Plomin)

n. The genetic changes in populations of *Drosophila pseudoobscura* in the american southwest (T. Dobzhansky)

o. The genetical evolution of social behavior (W.D. Hamilton)

p. Versuche über pflanzen-hybriden (G. Mendel)

Question 1-2. Weismann performed experiments in the late 1800's with inbred strains of mice possessing normal tails. He removed the tails of a group of mice and intermated individuals from this group. All of the offspring produced by parents with cut-off tails had normal length tails. This is a significant finding. Why? It shows that there is another reason other than physical appearances/mutations that determine the outcome of newly formed offspring.

Answers to General Questions

Q1-1. (a - g and p) Classical Genetics; (h - l) Molecular Genetics; and (m - o) Evolutionary Genetics.

Q1-2. This result is important because it showed that acquired characteristics are *not* inherited.

CHAPTER 2
MENDEL'S PRINCIPLES

Key Genetic Concepts

Variation: Regardless of the organism, individuals differ from one another in the characteristics they possess. In genetics we are concerned with two types of variation, **continuous** and **discrete**. Traits that vary continuously cannot be classified into a small number of distinct categories or classes. For example, consider height in human beings. We cannot easily place individuals in one of three categories: tall, medium, or short. We can however, order people from shortest to tallest with no clear distinctions between classes. Traits that vary continuously are usually influenced by environmental factors as well as many genes. In contrast, discrete traits are easily classified into several distinct categories. For example, consider Mendel's pea plants. Mendel studied plants that had white or purple flowers. All individuals could be placed into one of these categories and there were no individuals with intermediate appearances.

Phenotype and Genotype: As you begin to explore the field of genetics, you will find that organisms vary in their genetic makeup as well as their physical appearance. We use the term **phenotype** to describe the individual's physical attributes. The phenotype is "what can be seen." The term **genotype** refers to the genes that an organism possesses. Remember that a **gene** is the inherited determinant of the phenotype. Each individual usually receives two copies of each gene, one from the female parent and one from the male parent. The alternative forms of a gene are called **alleles**. For each gene, an individual may have two different alleles (**heterozygous** condition, a **heterozygote**) or two identical alleles (**homozygous** condition, a **homozygote**). The phenotype may be directly determined by the genotype or it may result from the combined effects of genes and the environment.

Mendelian Inheritance: Characteristics that are inherited in a simple Mendelian fashion are controlled by single genes with two alleles. Mendel focused on discrete traits with two alternative phenotypes in the garden pea. Mendel first established **pure-breeding** lines by self-fertilizing individual plants. These lines would "breed true" or produce only plants with the same phenotype. Mendel was able to cross these pure-breeding lines and observe the phenotypes of the offspring produced. He then counted and classified the resulting offspring according to phenotype. This **quantitative** approach differed from that traditionally used to investigate biological questions.

The pure-breeding individuals initially crossed represented the **parental** or **P1** generation. The offspring resulting from this cross comprised the **first filial** or **F1** generation. Individuals in the F1 generation are also referred to as **hybrids** because they are the result of a cross between two unlike parents. An **F2** generation is obtained by crossing individuals from the F1 generation. Mendel's crosses led to the conclusions outlined below. (Please note: Modern terminology such as "gene" and "allele" will be used in the discussion below for clarity. Mendel did not use these terms.)

Mendelian Concepts

Particulate Inheritance: Mendel's data clearly showed that physical particles were passed from parents to offspring. We now know that these particles are genes and that they are carried on the chromosomes. Each individual possess two alleles while the **gametes** (sperm and egg) contain only one.

Complete Dominance: While an individual has two alleles for each gene, both alleles are not necessarily expressed. Mendel consistently found that the F1 generation was composed of individuals of one phenotype while the P1 and F2 generations were composed of individuals with one of the two possible phenotypes. For example, crossing a plant with purple flowers to a plant with white flowers (P1) yields all purple flowered plants in the F1. A 3:1 ratio (purple:white) was found in the F2 generation. Mendel referred to the trait that was expressed in the F1 as **dominant**. The **recessive** trait was masked. Further crosses revealed that some of the F2 purple plants produced only purple offspring when self-fertilized while others produced a 3:1 **phenotypic ratio** of purple : white plants when they were self-fertilized. We now know that purple flowered plants in the F2 generation are either **homozygous dominant** (AA) or **heterozygous** (Aa) White plants produced only like offspring when selfed (aa, **homozygous recessive**). A 1:2:1 (AA : Aa : aa) **genotypic ratio** is observed in the F2 generation.

Rule of Segregation: During gamete formation, one allele from each allelic pair is randomly passed into each gamete. For example, an plant heterozygous for flower color (a **monohybrid**), is as likely to form gametes containing the allele for purple color as it is to form gametes containing the allele for white color.

Rule of Independent Assortment: While the rule of segregation focuses on the separation of alleles for a single gene, the rule of independent assortment considers the inheritance pattern of two or more genes simultaneously. When an individual heterozygous for two traits (AaBb; a **dihybrid**) forms gametes, the alleles for the first trait separate independently from those of the second trait. In this example, four different types of gametes will be formed in equal proportions: AB, Ab, aB, and ab.

Extensions of Mendelian Concepts

Incomplete Dominance: Dominance is not universal. Incomplete dominance occurs when one allele does not completely mask the expression of the second allele. Rather, both alleles influence the phenotype of a heterozygote resulting in an unique, intermediate expression. If two heterozygotes are crossed, the phenotypic and genotypic ratios are the same (1:2:1). For example, assume that coat color in cattle is determined by a pair of alleles that are incompletely dominant. If you mate a bull with red hair and a cow with white hair, the resulting offspring would be pink!

Codominance: Codominance is another type of inheritance pattern that results in a unique phenotype for the heterozygote. In this case, both alleles are expressed and the resulting phenotype is not an intermediate. Consider the cattle example above. In the case of codominance, the coat of the heterozygote would be composed of both red and white hairs (resulting in a roan coat color). If two heterozygotes are crossed, the phenotypic and genotypic ratios would again be the same (1:2:1).

Multiple Alleles: In the above examples, each gene has only two alleles. A gene can have more than two alleles although each individual still possesses only two of the possibilities (one from the egg and one from the sperm). One of the most familiar examples of a gene with multiple alleles is the human ABO blood system. In this system, the A and B alleles are dominant to the O allele but codominant with respect to each other. These relationships result in the four phenotypic classes observed (blood types A, B, AB, and O).

Epistasis: Epistasis is an example of a gene interaction in which two or more genes contribute to the same phenotype. In epistasis, one pair of alleles masks or modifies the expression of a second pair of alleles. Consider kernel color in corn. Kernels can either be either purple or white and this trait is controlled by two pairs of interacting genes. One pair of alleles (the first gene) determines if color is present or absent. The second pair of alleles (the second gene) determined the specific color (purple or white). An individual may have the genotype for purple color at gene B but be colorless (or white) because the genotype at gene A specifies no color production. The first gene is said to be **epistatic** to the second, or **hypostatic** gene.

One-Gene-One-Enzyme Hypothesis: This hypothesis focuses on the fact that gene expression results in the production of a protein, often an enzyme. These enzymes are important components of biosynthetic pathways. The one-gene-one-enzyme hypothesis has been modified to include the production of proteins that do not code for enzymes and is now referred to as the **one-gene-one-polypeptide hypothesis**.

Pleiotropy: In pleiotropy, a single gene has multiple effects on the phenotype. For example, sickle-cell anemia is a single gene defect that produces a wide variety of phenotypic symptoms.

Problem-Solving Hints

Hint 1: Read the question carefully and determine what information is requested. For example, a question may ask for the *genotypic* ratio among offspring from a specific cross while another question may ask for the phenotypic ratio expected. In some cases, you may be provided with a set of parents to cross and asked to provide information about the offspring characteristics. Alternatively, the question may request information about the parental genotype/phenotype.

Hint 2: If appropriate, translate phenotypes into symbols and diagram, insofar as possible, the cross under consideration. Use the symbols you establish consistently throughout the problem. Remember that each individual is diploid and therefore must have two alleles for each trait. Eggs and sperm normally contain only one allele for each trait. In cases involving the inheritance of two or more traits, always group alleles for the same trait together.

Hint 3: To determine the pattern of inheritance, examine the phenotypes and genotypes of each generation.

a. The presence of two alternative phenotypes *suggests* complete dominance and a single gene.

b. The presence of an intermediate phenotype in the F1 generation *suggests* incomplete dominance. The expression of both alleles in a heterozygote *suggests* codominance.

c. Consider the *possibility* of multiple alleles if identical crosses produce different phenotypes. For example, examine the offspring produced from the mating of two parents with type A blood. In the first cross, all of the offspring have type A blood. In the second cross, three individuals have type A blood while one individual has type O blood.

d. Epistasis is *indicated* when the phenotypic ratio among offspring is expressed as a modified 9:3:3:1 ratio and there are typically fewer than four phenotypic classes.

Hint 4: Examine the phenotypic ratio of the progeny to determine the genotype of the parents.

a. A 3:1 ratio among the offspring *suggests* that both parents are heterozygous.

b. A 1:1 ratio among the offspring *suggests* that one parent is heterozygous and the other is homozygous recessive.

c. A 1:2:1 ratio among the offspring *suggests* that both parents are heterozygous and the mode of inheritance is incomplete dominance or codominance.

d. A 9:7, 9:3:4, 12:3:1, 13:3, or 15:1 ratio *suggests* epistasis.

e. A 2:1 ratio *suggests* a homozygous lethal and that both parents are heterozygous.

f. A 9:3:3:1 ratio *suggests* that the parents are heterozygous for two traits.

g. A 1:1:1:1 ratio *suggests* that, for each of two traits, one of the parents is heterozygous while the other parent is homozygous recessive. There are two possible crosses: AaBb x aabb and Aabb x aaBb.

h. A 3:3:1:1 ratio *suggests* that one parent is heterozygous for two traits while the other parent is heterozygous for one gene and homozygous recessive for the second gene.

Hint 5: Each reproductive event is independent. The ratios given above are predicted given large numbers of offspring. For example, do not mistakenly believe that a couple (each carrying a recessive disease allele) will necessarily have three normal children if their first child was affected. Rather, remember that each time the couple conceives a child, there is a 25% chance that the child will be affected regardless of the outcome of earlier pregnancies.

Practice Problems

Problem 2-1. Consider the condition *epicanthus* in humans. This disorder, a folding of the upper eyelid, is inherited as an autosomal dominant. An affected male marries a normal woman and they start a family. What proportion of their children are expected to be affected by this disorder? What are the genotypes of the man and his wife?

Problem 2-2. The offspring of a dihybrid cross are individually testcrossed. Predict the frequency of progeny expected in each phenotypic class.

Problem 2-3. A cross is made between a bird with blue feathers and a bird with yellow feathers. All of the offspring have green feathers. Two of the birds with green feathers are crossed and the resulting generation consists of 10 blue birds, 20 green birds, and 9 yellow birds. How is feather color inherited? If one of the green birds is backcrossed to the yellow parent, what genotypes and phenotypes are expected in their progeny?

Problem 2-4. For each of the following couples, determine the possible genotypes and phenotypes of their children:

a. type A male and type A female

b. type B male and type A female

c. type AB male and type AB female

d. type AB male and type O female

e. type O male and type O female

Problem 2-5. The Bombay phenotype in humans is a rare recessive mutation. Normally, the gene products of the A and B alleles of the ABO blood system modify the terminal sugars of a mucopolysaccharide (H structure) of the red blood cells. Individuals with the genotype HH or Hh have a complete H structure. However, individuals with the hh genotype produce an H substance lacking a fructose. The enzymes specified by the A and B alleles of the ABO blood system are unable to recognize and modify this incomplete H substance. As a result, the ABO system genotype is not expressed in an individual that is homozygous for the H substance abnormality (hh). These individuals are phenotypically type O. Predict the phenotypic ratio found in the progeny of the following couple: I^AI^B, Hh x I^AI^B Hh.

Problem 2-6. In mice, the allele for short tail (A) is dominant over the allele for long tail (a). The genotype AA is lethal. Predict the results of a cross between two mice with short tails.

Problem 2-7. Consider the crosses in garden peas provided in the table below. For each of the two traits (flower color and pod position), determine which is dominant and which is recessive. For each cross, determine the genotype of both parents.

	Offspring Phenotypes			
Cross	**Purple Axial**	**Purple Terminal**	**White Axial**	**White Terminal**
Purple, Axial x Purple, Axial	99	32	29	12
White, Axial x Purple, Axial	112	38	114	40
Purple, Terminal x White, Axial	150	0	155	0
Purple, Terminal x White, Axial	100	99	101	104
White, Terminal x White, Terminal	0	0	0	192

General Questions

Question 2-1. Describe how Mendel's findings about inheritance differed from the ideas of his contemporaries.

Question 2-2. Mendel's work was published in 1866 but was virtually ignored until the early 1900's. Why did the scientific community "overlook" this major contribution? Who rediscovered Mendel's principles?

Question 2-3. Would it be possible to demonstrate independent assortment using a monohybrid cross? Explain.

Question 2-4. Compare and contrast complete dominance, incomplete dominance, and codominance.

Question 2-5. Describe how genes control the fate of metabolic pathways.

Solutions to Practice Problems

P2-1. The trait under consideration is inherited as an autosomal dominant; therefore both homozygotes and heterozygotes will be affected. The male parent in this problem is affected. His genotype is either EE or Ee. The female parent is phenotypically normal and therefore has the genotype ee. To answer this question, two different crosses must be performed because the specific genotype of the male is unknown. If the male is homozygous dominant, then all of the offspring produced would have the genotype Ee and would be affected. If the male is a heterozygote, one half of his sperm will contain the dominant allele (E) and the other half of the sperm will carry the normal allele (e). All of the female's eggs contain the normal allele (e). Offspring resulting from this cross will have a 50% chance of expressing the condition (genotype Ee) and a 50% chance of normality (genotype ee).

P2-2. In the case of a testcross, the individual with an unknown genotype is crossed with a homozygous recessive individual. The genotype of the unknown individual will ultimately determine the phenotype observed: a dominant phenotype can only result if the unknown individual contributes a dominant allele.

Parent Genotype	Gametes Produced
AABB	all AB
AaBB	½ AB and ½ aB
AABb	½ AB and ½ Ab
AaBb	¼ AB, ¼ Ab, ¼ aB, and ¼ ab
AAbb	all Ab
Aabb	½ Ab and ½ ab
aaBB	all aB
aaBb	½ aB and ½ ab
aabb	all ab

Testcrosses	Frequency of each Phenotype: Dominant A Dominant B	Dominant A Recessive B	Recessive A Dominant B	Recessive A Recessive B
AABB x aabb	100%	0	0	0
AaBB x aabb	50%	0	50%	0
AABb x aabb	50%	50%	0	0
AaBb x aabb	25%	25%	25%	25%
AAbb x aabb	0	100%	0	0
Aabb x aabb	0	50%	0	50%
aaBB x aabb	0	0	100%	0
aaBb x aabb	0	0	50%	50%
aabb x aabb	0	0	0	100%

P2-3. Begin by examining the phenotypic classes and ratios produced in the F1 and F2 generations. The F1 offspring were all green, a color intermediate to the parental colors of blue and yellow. In the F2 generation, three distinct phenotypes are observed in a 1:2:1 ratio. One half of the offspring show the intermediate phenotype (green) of the F1 generation. This pattern indicates that feather color shows incomplete dominance (*Hint 3b*). The green birds are heterozygotes. Crossing a green bird (Aa) with the yellow parent (aa) will yield a 1:1 ratio of green (Aa): yellow (aa) birds.

P2-4. The ABO blood system displays complete dominance as well as codominance. The A and B alleles are dominant to the O allele but are codominant with respect to each other. Before you begin to cross the individuals below, determine the possible genotypes corresponding to each blood type. Blood type A can result from genotype I^AI^A or I^AI^O; blood type B can result from genotype I^BI^B or I^BI^O; blood type AB corresponds to genotype I^AI^B; and blood type O corresponds to genotype I^OI^O.

Cross	Offspring Genotypes	Offspring Phenotypes
cross a		
I^AI^A x I^AI^A	all I^AI^A	all type A
I^AI^A x I^AI^O	½ I^AI^A, ½ I^AI^O	all type A
I^AI^O x I^AI^O	¼ I^AI^A, ½ I^AI^O, ¼ I^OI^O	¾ type A, ¼ type O
cross b		
I^BI^B x I^AI^A	all I^AI^B	all type AB
I^BI^O x I^AI^A	½ I^AI^B, ½ I^AI^O	½ type AB, ½ type A
I^BI^B x I^AI^O	½ I^AI^B, ½ I^BI^O	½ type AB, ½ type B
I^BI^O x I^AI^O	¼ I^AI^B, ¼ I^BI^O, ¼ I^AI^O, ¼ I^OI^O	¼ type AB, ¼ type B, ¼ type A, ¼ type O
cross c		
I^AI^B x I^AI^B	¼ I^AI^A, ½ I^AI^B, ¼ I^BI^B	¼ type A, ½ type AB, ¼ type B
cross d		
I^AI^B x I^OI^O	½ I^AI^O, ½ I^BI^O	½ type A, ½ type B
cross e		
I^OI^O x I^OI^O	all I^OI^O	all type O

P2-5. The Bombay phenotype is an example of epistasis. There are two gene pairs involved. One pair of alleles specifies the blood type. The second pair of alleles masks the expression of the first gene by interfering with the production of a complete H substance. As a result, the blood system genotype is only expressed in the presence of a dominant H allele at the second locus. The cross you are asked to consider is a dihybrid cross: both individuals are heterozygous for both genes. The genotypes resulting from a dihybrid cross are easily predicted. The phenotypes associated with each genotype will be dependent upon the interaction of the two genes. The final phenotypic ratio observed is: 3/16 type A, 3/16 type B, 6/16 type AB, and 4/16 type O.

Combined Genotype	ABO Phenotype	Complete H?	Expressed Phenotype
1/16 I^AI^A HH	Type A	YES	1/16 Type A
2/16 I^AI^A Hh	Type A	YES	2/16 Type A
1/16 I^AI^A hh	Type A	NO	1/16 Type O
2/16 I^AI^B HH	Type AB	YES	2/16 Type AB
4/16 I^AI^B Hh	Type AB	YES	4/16 Type AB
2/16 I^AI^B hh	Type AB	NO	2/16 Type O
1/16 I^BI^B HH	Type B	YES	1/16 Type B
2/16 I^BI^B Hh	Type B	YES	2/16 Type B
1/16 I^BI^B hh	Type B	NO	1/16 Type O

P2-6. All mice with short tails are heterozygous because the AA genotype is lethal. If you cross two heterozygotes, you expect to find a ¼ AA : ½ Aa : ¼ aa genotypic ratio. Elimination of the AA individuals yields 2/3 short tailed mice and 1/3 long tailed mice (*Hint 4e*).

P2-7. To begin this problem, you need to first examine all crosses and determine which traits are dominant. Notice that the purple, axial plants in cross (a) produced a generation of plants containing all possible combinations of traits in approximately a 9:3:3:1 ratio. This ratio is expected from a cross of two dihybrids (*Hint 4f*). The traits possessed by the dihybrids (purple and axial) are dominant. Once the dominant and recessive traits have been identified, fill in as many of the genotypes as possible (e.g. you know that white plants are genotype pp while purple plants have at least one dominant allele P). To determine if plants with the dominant phenotype are homozygous or heterozygous, examine the phenotypes produced as well as the phenotypic ratios. For example, we can deduce that the purple plant in (b) is heterozygous because the cross produced a 1:1 ratio of plants with purple flowers to plants with white flowers (*Hint 4b*). Genotypes for (b) could also be assigned by recognizing the 3:3:1:1 ratio (*Hint 4h*).

a.	PpAa x PpAa	9:3:3:1
b.	ppAa x PpAa	1 purple : 1 white; 3 Axial : 1 Terminal
c.	Ppaa x ppAA	1 purple : 1 white; all axial
d.	Ppaa x ppAa	1:1:1:1
e.	ppaa x ppaa	all white terminal

Answers to General Questions

Q2-1. Mendel's differed from his contemporaries primarily in his approach to the question of inheritance. Mendel used a quantitative, rather than qualitative, approach. He also conducted simple experiments and looked for statistical regularities (e.g. a 3:1 ratio).

Q2-2. The reason why Mendel's work was "overlooked" by the scientific community is a point of contention. Most believe that his use of mathematics to explain biological phenomenon isolated many of his contemporaries. His work was independently rediscovered by Correns, Tschermak, and de Vries.

Q2-3. A monohybrid cross can not be used to demonstrate independent assortment. Independent assortment states that segregation of a pair of alleles is independent of the separation of *other* pairs. In a monohybrid cross, only one pair of alleles is under consideration.

Q2-4. In complete dominance, one allele is completely expressed while the other is masked in a heterozygous individual. Both homozygous dominant and heterozygous genotypes produce a dominant phenotype. In contrast, the heterozygote has its own phenotype in incomplete dominance and codominance. The heterozygous phenotype is intermediate to the parental phenotypes if the trait is incompletely dominant. In codominance, the heterozygous phenotype results from the complete expression of both alleles.

Q2-5. Genes produce proteins. Many of these proteins are enzymes which participate in metabolic pathways. An abnormal gene may fail to produce a necessary enzyme thus shutting down a specific metabolic pathway or part of the pathway.

Answers to Even Numbered End of Chapter Exercises, Problems and Critical Thinking Questions

2. Properties that make an organism popular among geneticists include a well-understood life cycle; ease of maintenance in the laboratory or in experimental gardens; a lot of variability in the phenotype that is genetically controlled; ease of manipulation (breeding); and short generation time.

4. *Aa* × *aa*. We see two phenotypes in approximately a 1:1 ratio in the offspring. We don't know which allele is dominant from this cross. The two phenotypes indicate one gene, and the 1:1 ratio indicates a mating of a homozygote and a heterozygote (the testcrossing of a heterozygote).

6. Red eye is dominant, and the cross is between two heterozygotes since we see two phenotypes in an approximate 3:1 ratio in the offspring.

8. The *A* and *B* alleles produce glycosyl transferase enzymes that each add a different substance to the H structure, whereas the *O* allele produces a functionless protein that adds nothing. Thus, for example, the *AO* heterozygote converts all the H structures to the *A* antigen making *A* dominant to *O*. *AB* heterozygotes convert the H structures to approximately equal numbers of A and B antigens.

10. Pink seems to be a blend of red and white. Assume that the red and white plants are homozygtes. If one of the colors were dominant, the progeny from the first cross should all be either red or white. This is not seen; therefore, neither color is dominant.

Pink probably represents the heterozygote. Let c = the gene color: c^r = red and c^W = white. The first cross can be indicated:

$$c^r c^r \times c^W c^W$$

(red) (white)

↓

all $c^r c^W$

(pink)

Therefore, the second cross is:

$$c^r c^W \times c^r c^W$$

(pink) (pink)

Any two heterozygotes will always yield a 1:2:1 genotypic ratio, or, in case, ¼ $c^r c^r$: ½ $c^r c^W$: ¼ $c^W c^W$. Approximately half of the progeny should be pink. We examine the cross and see that 23 out of 46 are pink.

12. In the table: D, dominant; R, recessive; M, mutant; +, wild-type.

Allele	*Dominant or Recessive*	*Mutant or Wild-Type*	*Alternative Allele*	*Dominant or Recessive*
y^+	D	+	y	R
Hw	D	M	Hw^+	R
Ax^+	R	+	Ax	D
Co	D	M	Co^+	R
rv^+	D	+	rv	R
dow	R	M	dow^+	D
$M(2)e^+$	R	+	$M(2)e$	D
J	D	M	J^+	R
tuf^+	D	+	tuf	R
bur	R	M	bur^+	D

14. The following table shows female and child phenotypes in the ABO system and the phenotypes of men who cannot be the father of the child.

Mother	*Child*	*Nonfathers*	*Mother*	*Child*	*Nonfathers*
A	A	—	AB	A	—
	B	A, O		B	—
	AB	A, O		AB	O
	O	AB			
B	A	B, O	O	A	B, O
	B	—		B	A, O
	AB	B, O		O	AB
	O	AB			

16. Constraints are set by the length of the gene (number of mutable sites) and the different possible phenotypic effects that can result from alteration of the amino acid sequence of the protein product of the gene.

18. In each cross we see that all ratios could result from a single gene (3:1, 1:2:1, and 1:1). We can easily explain the red and silver if we assume that red is dominant over silver. Thus, the first cross could be:

$$RR \times rr \text{ or } Rr \times rr$$

Similarly, the second cross could be :

$$RR \times R- \text{ or } Rr \times Rr$$

But we cannot explain any of the white progeny with only two alleles. Since all ratios are single gene ratios, we must propose a third allele that, when homozygous, produces white. Red must be dominant to white. Note that a given diploid individual can have only two alleles. To list all the possible genotypes and their phenotypes, let R = red and r = nonred.

$$RR, Rr^s, Rr^w \text{ : red (dominant to all other alleles)}$$

$$r^s r^s \text{ : silver}$$

$$r^w r^w \text{ : white}$$

What is the phenotype of an $r^s r^w$ individual? Assume silver is dominant; that is, assume r^s is dominant to r^w.

Therefore $r^s r^w$ = silver.

Now go back to the results. By our model, $r^w r^w$ is white. Therefore, in order to get white progeny, each parent must have at least one r^w allele to contribute. The cross of red by silver to produce white offspring must be:

$$Rr^w \times r^s r^w$$

$$1/4Rr^s : 1/4Rr^w : 1/4r^s r^w : 1/4r^w r^w$$

(red) (red) (silver) (white)

20. The round, yellow F_2 plants are made up of four genotypes; the round, green of two genotypes; the wrinkled, yellow of two genotypes; and the wrinkled, green of one genotype. Testcrossing all these genotypes produces the following results:

$R\text{-}Y\text{-}:$ $1/16\,RRYY \times rryy \rightarrow$ all $RrYy$

$2/16\,RrYY \times rryy \rightarrow 1/2\,RrYy, 1/2\,rrYy$

$2/16\,RRYy \times rryy \rightarrow 1/2\,RrYy, 1/2\,Rryy$

$4/16\,RrYy \times rryy \rightarrow 1/4\,RrYy, 1/4\,Rryy, 1/4\,rrYy, 1/4\,rryy$

$R\text{-}yy:$ $1/16\,RRyy \times rryy \rightarrow$ all $Rryy$

$2/16\,Rryy \times rryy \rightarrow 1/2\,Rryy, 1/2\,rryy$

$rrY\text{-}:$ $1/16\,rrYY \times rryy \rightarrow$ all $rrYy$

$2/16\,rrYy \times rryy \rightarrow 1/2\,rrYy, 1/2\,rryy$

$rryy:$ $1/16\,rryy \times rryy \rightarrow$ all $rryy$

22. $CcTt \times$ self yields:

$3/16\,C\text{-}TT$ purple, tall

$6/16\,C\text{-}Tt$ purple, medium

$3/16\,C\text{-}tt$ purple, dwarf

$1/16\,ccTT$ white, tall

$2/16\,ccTt$ white, medium

$1/16\,cctt$ white, dwarf

24. A:AB:B is 4:8:4 or 1:2:1; Rh^+:Rh^- is 12:4 or 3:1. Both loci are segregating properly. The lack of independent assortment would result in the lack of simple ratios (see chapter 6).

26. When a decahybrid is selfed, it would produce 2^{10} = 1,024 different gametes; $1/(2^{10})^2 = 9.53 \times 10^{-7}$ of the F_2 would be homozygous recessive; 3^{10} = 59,049 different genotypes yielding 2^{10} different phenotypes would appear. If the decahybrid were testcrossed, it would produce 2^{10} different gametes; $1/2^{10} = 0.00098$ of the F^2 would be homozygous recessive; 2^{10} different genotypes and phenotypes would appear.

28. Examine each trait separately. There are 104 long:34 short; and 69 brown:69 red. Length appears in a 3:1 ratio; therefore, long is dominant and each parent is heterozygous. Eye color appears in a 1:1 ratio. We can't conclude which allele is dominant; all we can conclude is that one parent is a recessive homozygote and that one parent is a heterozygote. (We will assume that red is the wild-type.) If s^+ = long, s = short, bw^+ = red, and bw = brown, one possible way to indicate the cross is $s^+s\ bwbw \times s^+s\ bw^+bw$.

30. You could set up a Punnett square and count the boxes; but unfortunately, this is an 8 × 8 matrix that yields 64 squares to count. Very tedious. Set up the cross:

$$Tt\ Yy\ Rr \times Tt\ Yy\ Rr$$

and look at one gene at a time.

a. The chance of getting each dominant trait is 3/4. Therefore, the chance of all three together is 3/4 × 3/4 × 3/4 = 27/64.

b. The chance of getting each recessive trait is 1/4. Therefore, the chance for all three recessives is 1/4 × 1/4 × 1/4 = 1/64.

c. The chance of short is 1/4, the chance of green is 1/4, and the chance of round is 3/4. Therefore the total chance is 1/4 × 1/4 × 3/4 = 3/64.

32. Red and crossveins are dominant.

a. $oror\ cv^+cv \times oror\ cv^+cv$

b. $or^+or\ cv^+cv \times or^+or\ cvcv$

c. $or^+\text{-}\ cvcv \times or^+or^+\ cv^+cv$

d. $or^+or\ cv^+cv \times or^+or\ cv^+cv$

Let or^+ = red, or = orange, cv^+ = crossveins, cv = crossveinless. Look first at those crosses that yield only two phenotypes and thus indicate that one gene is homozygous. Crosses a and c are such crosses. In cross a we see only orange eyes, and in cross c we see only red eyes. The eye color must be homozygous in at least one parent. From these two crosses we can't determine whether red or orange is dominant. In cross a, both parents have crossveins, and we see a 3:1 ratio in the progeny. Crossveins must be dominant, and this cross must be $cv^+cv \times cv^+cv$. Similar logic applies to cross c.

In cross b we see an approximate 3:3:1:1 ratio, which suggests that the same gene is heterozygous in each parent. We see a 3:1 red:orange ratio. Therefore, red is dominant and heterozygous in each parent: $or^+or \times or^+or$. We have already determined that the allele for crossveins is dominant, which indicates the cross $or^+or\ cv^+cv \times or^+or\ cvcv$. Thus a 3:3:1:1 ratio indicates this general situation: $AaBb \times Aabb$. In cross d we see four phenotypes in an approximate 9:3:3:1. This result indicates a mating between two double heterozygotes $or^+or\ cv^+cv \times or^+or\ cv^+cv$.

34.

a. 1/16 or 4/64

b. 33/64

To be colorless, individuals must be at least *aabb*; they could be either *aa bb C-* or *aa bb cc*. The chance of the first genotype is (1/4 *aa*)(1/4 *bb*)(3/4 *C-*) = 3/64, and the chance of the second is (1/4 *aa*)(1/4 *bb*)(1/4 *cc*) = 1/64. Since both phenotypes are colorless, we add probabilities to get the total, 4/64. Or, (1/4 *aa*)(1/4 *bb*)(1 *C-* or *cc*) = 1/16.

For red probabilities, determine genotypes that will be red, calculate probability for each genotype, and add.

$$A\text{-}\ bb\ C\text{-}\ 3/4 \times 1/4 \times 3/4 = 9/64$$
$$A\text{-}\ bb\ cc\ 3/4 \times 1/4 \times 1/4 = 3/64$$
$$A\text{-}\ B\text{-}\ cc\ 3/4 \times 3/4 \times 1/4 = 9/64$$
$$aa\ B\text{-}\ C\text{-}\ 1/4 \times 3/4 \times 3/4 = 9/64$$
$$aa\ B\text{-}\ cc\ 1/4 \times 3/4 \times 1/4 = \underline{3/64}$$
$$33/64$$

Alternatively, calculate the frequency of blacks as 3/4 *A-* × 3/4 *B-* × 3/4 *C-* = 27/64. We calculated 4/64 colorless, so 1 – (27/64 + 4/64) = 33/64 = frequency of red.

36. 119:32:9 is very close to a 12:3:1 ratio indicating two loci with epistasis. For example:

P_1 red (*RRYY*) × white (*rryy*)

F_1 red (*RrYy*) × self

F_2 119 (*R-Y-* + *R-yy*):32 (*rrY-*):9 (*rryy*)

38. The animals would be bred for several generations of controlled matings. You would look for offspring types and ratios consistent with a hypothesis of the number of genes controlling the trait and the dominance relationships of alleles. One approach might be to try to establish pure-breeding lines (homozygotes), which can then be bred in a more controlled manner. If both rooster and hen were pure-breeding rose-combed types (all offspring were also rose-combed), then it would not be possible to determine the nature of the genetic control of this trait.

40. Dominance refers to the expression of an allele (and the masking of the other) when in the heterozygous condition. Epistasis is similar to dominance but is intergenic; that is, an allelic combination at one gene can mask the expression of alleles of another gene. Pleiotropy is suspected when different phenotypic effects are observed together. It is supported when crosses cannot separate these effects and verified when a molecular or physiological mechanism of gene action is determined.

42.

Blockage	*Build Up*	*Never Appears*
1	Q	R, S, T, U
2	R	S, T, U
3	S	T, U
4	T	U

44.

$$\begin{array}{ccccccccccc} & & 4 & & 2 & & 5 & & 3 & & 1 \\ ? & \rightarrow & D & \rightarrow & B & \rightarrow & C & \rightarrow & A & \rightarrow & E \end{array}$$

Accumulation :1, A; 2, D; 3, C; 4, ?; 5, B

Critical Thinking Question

2. If Mendel self-fertilized plants that were heterozygous for all seven loci *(AaBbCcDdEeFfGg)*, he could study it with a Punnett square with 16,384 squares. This is arrived at by noting that since each locus can produce two types of gametes (*A* and *a*), the total number of different gametes of the septa-hybrids is 2^7 (2 × 2 × 2 × 2 × 2 × 2 × 2) = 128. That number squared is the number of boxes in the Punnett square. The easiest way to answer questions about the offspring ratios is with simple probabilities, discussed in more detail in chapter 4. Thus, for any locus, the probability of the dominant homozygote is ¼; the heterozygote is ½; and the recessive homozygote is ¼. We can thus answer questions such as: What is the probability of an F_2 offspring having the genotype of *AABBCcDdEEFFgg?* For this we will multiply probabilities at each locus: ¼ × ¼ × ½ × ½ x ¼ × ¼ × ¼, or $1/(4^5 \times 2^2)$ = 1/4096 = 0.0002.

CHAPTER 3
MITOSIS AND MEIOSIS

Key Genetic Concepts

Eukaryotic Chromosome Structure: Chromosomes are structures carrying genes. After replication, chromosomes are composed of two genetically identical strands (**sister chromatids**) held together by a single **centromere**. The DNA is complexed with protein and the chromosomes are linear arrangements of genes. Eukaryotic chromosomes are classified according to the position of the centromere: in the middle (**metacentric**), very near the end (**acrocentric**), between the middle and the end (**subtelocentric/submetacentric**), or at the end (**telocentric**) of the chromosome structure.

Chromosome Number (Ploidy): Most eukaryotic cells are **diploid** (**2N**). In a diploid cell, each type of chromosome is represented twice in the nucleus. One set of chromosomes is maternally derived while the second set is paternally derived. Each pair of morphologically similar chromosomes is referred to as a **homologous pair**. **Gametes**, sperm and egg, usually contain only one set of chromosomes and are referred to as **haploid** (**1N**).

The Cell Cycle and Mitosis: Most cells undergo a continuous cycle of growth, DNA synthesis, and division. The cell cycle is divided into two main stages, interphase and mitosis. During interphase, the cell grows (**Gap 1** or **G1** phase), synthesizes DNA (**Synthesis** or **S** phase), and continues to prepare to divide (**Gap 2** or **G2** phase). Following this progression of phases (G1 to S to G2), the cell divides during the processes of mitosis and cytokinesis. **Mitosis** is a nuclear division in which a diploid cell gives rise to two identical diploid cells. Mitosis, while not a lengthy process compared to the entire cell cycle, is further subdivided into phases: **prophase**, **metaphase**, **anaphase**, and **telophase**. During mitosis, the sister chromatids of each chromosome are separated into two different cells. In this way, each resulting cell receives an identical set of chromosomes. The **spindle apparatus** is responsible for the movement of chromosomes during mitosis. The nuclear division (**karyokinesis**) is followed by a division of the cytoplasm (**cytokinesis**). The progression of the cell cycle is genetically controlled.

Meiosis: Meiosis is a nuclear division in which a diploid cell gives rise to four haploid cells. Meiosis is a reduction division in which the number of chromosomes is halved. There are two successive cell divisions: **meiosis I** and **meiosis II**. In meiosis I, homologous pairs of chromosomes are separated resulting in two *haploid* cells. This division is referred to as the reduction division. In meiosis II, sister chromatids are separated giving rise to four haploid cells. The second meiotic division equalizes the DNA content of the cells produced. Both meiosis I and II are subdivided into phases: prophase, metaphase, anaphase, and telophase.

Meiosis in Animals: In animals, meiosis occurs in the gonads and results in the production of sperm or eggs. **Spermatogenesis**, occurring in male animals, produces four equally sized **sperm cells**. In females, **oogenesis** results in the formation of four haploid cells of unequal size: a larger **ovum** (egg) and three smaller polar bodies. Meiosis is an integral component of sexual reproduction in animals for two reasons. First, meiosis provides a mechanism for the maintenance of chromosome number. Meiosis produces haploid gametes which, at fertilization, restore the diploid number. Second, meiosis generates variation through the random distribution of maternally and paternally derived chromosomes and through the process of **crossing over (recombination)**.

Life Cycles: In most eukaryotic organisms, an alternation between a diploid and a haploid state is observed. In plants, there is an alternation of generations with important roles for both mitosis and meiosis. The flowering plant is a diploid **sporophyte**. The sporophyte gives rise to a haploid **gametophyte** through the process of meiosis. The gametophyte produces gametes by mitosis.

Chromosomal Theory of Inheritance: This theory was proposed by Walter Sutton in 1902. The theory states that chromosomes are linear sequences of genes. The behavior of chromosomes during meiosis correlates with, and provides an explanation for, Mendel's principles.

Problem-Solving Hints

Hint 1: To determine the number of chromosomes in a cell, count the number of centromeres. One centromere equals one chromosome.

Hint 2: The amount of DNA in a cell is doubled prior to the beginning of mitosis and meiosis. In mitosis, the DNA is divided equally between the two resulting cells. In meiosis, the DNA is divided equally among the four cells produced. Although there are two divisions during meiosis, DNA synthesis does not occur between divisions.

Hint 3: The terms diploid and haploid refer to the number of *chromosome sets*. A diploid number of four (2N=4) indicates that there are four chromosomes (two pairs) in the cell. A haploid number of four (N=4) indicates that there are four, unpaired chromosomes in the cell.

Hint 4. To determine which division process is taking place, examine the chromosome complement. In mitosis, homologous pairs of chromosomes are present. During mitosis, these chromosomes align singly along the metaphase plate. During meiosis I, homologous pairs of chromosomes are present but they align paired on either side of the metaphase plate. In meiosis II, the cell contains half the original number of chromosomes (only one member of each homologous pair is present).

Practice Problems

Problem 3-1. Consider a diploid cell with 2N=24. Assume that the amount of DNA in this cell at G1 of interphase is represented by the quantity "*X*." Complete the table below.

Stage of Division	Number of Chromosomes	Number of Chromatids	Quantity of DNA
At Prophase of Mitosis	24	48	2X
After Completion of Mitosis	24	0	X
At Prophase of Meiosis I	24	48	2X
At Prophase of Meiosis II	12	24	X
In Sperm and Egg Cells	12	0	X/2

Problem 3-2. *Drosophila melanogaster* has four pairs of chromosomes in the diploid state. Diagram the chromosome complement of a cell from an adult fruit fly and from a fruit fly gamete.

Problem 3-3. Consider a cell containing a pair of metacentric chromosomes and a pair of telocentric chromosomes. Diagram the chromosome composition expected to result from a (a) mitotic division, (b) meiosis I, and (c) meiosis II.

Problem 3-4: How many combinations of maternal and paternal chromosomes are possible in the gametes if the diploid number of chromosomes in the parent cell is (a) 2; (b) 4; or (c) 6?

Problem 3-5. If purple endosperm (P) is dominant to white (p) in corn, give the genotype and phenotype of the endosperm cells that would result by transferring pollen from a homozygous PP plant to the pistils of a homozygous pp plant. If the pollinating plant were pp and the pistillate plant PP, would the phenotypic results be different?

General Questions

Question 3-1. Describe the main features of prophase, metaphase, and anaphase of mitosis, meiosis I, and meiosis II.

Question 3-2. Describe how the processes of mitosis and meiosis explain Mendel's principles of segregation and independent assortment.

Question 3-3. Would the sequence of genes found on one of the sister chromatids of a dyad necessarily be identical to the sequence of genes found on a bivalent occurring earlier in meiosis? Explain.

Question 3-4. Meiosis is often termed a "reduction division." What is reduced and when?

Question 3-5. Given that most mutations occur during DNA synthesis, will sperm cells or egg cells be expected to carry more new mutations? Explain.

Question 3-6. Compare the processes of mitosis and meiosis with regard to each of the following features:

a. Number of cell divisions
b. Ploidy and number of resulting products
c. Number of times DNA replicates
d. Synapsis of homologous chromosomes
e. Disjunction of homologous chromosomes
f. Ability of products to divide by meiosis
g. Function and significance

Solutions to Practice Problems

P3-1. The cell in question has 24 chromosomes (12 pairs) in the nucleus (Hint 3). During the division process, each chromosome is composed of two chromatids joined by a centromere. If the amount of DNA present in the cell at G1 of interphase is "*X*," the amount of DNA present in the cell prior to mitosis or meiosis will be "2*X*" (*Hint 2*).

Stage of Division	Number of Chromosomes	Number of Chromatids	Quantity of DNA
At Prophase of Mitosis	24	48	2*X*
After Completion of Mitosis	24	0	*X*
At Prophase of Meiosis I	24	48	2*X*
At Prophase of Meiosis II	12	24	*X*
In Sperm and Egg Cells	12	0	*X*/2

P3-2. Cell from adult Gamete

P3-3. Assume that the two circles in a-c (below) represent the two cells resulting from a single division.

a. Mitosis b. Meiosis I c. Meiosis II

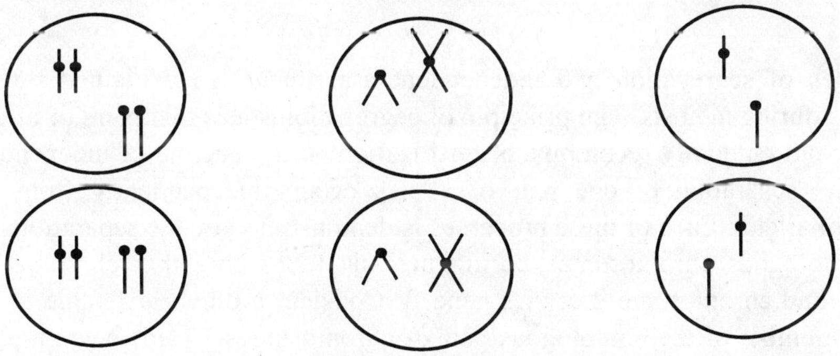

P3-4. To determine the number of gametes possible, use the equation 2^n where n is the number of chromosome pairs.

a. $2^2 = 2 \times 2 = 4$

b. $2^4 = 2 \times 2 \times 2 \times 2 = 16$

c. $2^6 = 2 \times 2 \times 2 \times 2 \times 2 \times 2 = 64$

P3-5. Examination of the life cycle of corn reveals that endosperm in triploid (3N). This ploidy level results from the fusion of two polar nuclei and one sperm nucleus (see Tamarin Figure 3.30). Therefore the endosperm genotype will be Ppp. If the female parent were genotype PP, the resulting endosperm genotype will be PPp. Both genotypes produce purple kernels.

Answers to General Questions

Q3-1.

Mitosis	Prophase	Nuclear membrane disintegrates, spindle apparatus forms, chromosomes condense
	Metaphase	Chromosomes align with centromeres on equator
	Anaphase	Chromatids separate and move to opposite poles
Meiosis I	Prophase	Nuclear membrance disintegrates, Sspindle apparatus forms, chromosomes condense, homologous chromosomes synapse
	Metaphase	Homologous chromosomes align on each side of the equator
	Anaphase	Homologous chromosomes separate and move to opposite poles
Meiosis II	Prophase	Chromosomes contract
	Metaphase	Chromosomes align with centromeres on equator
	Anaphase	Chromatids separate and move to opposite poles

Q3-2. Mendel's principles of segregation and independent assortment are explained by the behavior of chromosomes during meiosis. The principle of segregation states that pairs of alleles separate from each other and randomly recombine at fertilization. The principle of independent assortment states that the separation of one pair of alleles occurs independently from the separation of other pairs of alleles. Each of these processes is demonstrated by the separation and varied alignment of homologous chromosomes during meiosis I. Assume that chromosome number 1 carries gene A and chromosome 2 carries gene B. Consider a dihybrid female in the process of meiosis. One member of the homologous pair of chromosomes 1 (1m) bears the A1 allele. The other member of the pair (1p) contains the A2 allele. At anaphase I, the homologous chromosomes disjoin. As a result, A1 is partitioned into one cell while A2 appears in the other cell. To demonstrate independent assortment, we must simultaneously consider genes A and B. Because chromosomes 1 and 2 separate from each other independently, so do the alleles of genes A and B. As a result, a gamete may contain any allele combination from the A and B genes (refer to figure below). If the cell received chromosome 1m and 2m, the genotype would be A1B1. If the cell received chromosome 1m and 2p, the genotype would be A1B2. A chromosome complement of 1p and 2p results in the A2B2 genotype. Finally, a gamete would have the genotype A2B1 if it contained chromosomes 1p and 2m.

Dihybrid female:

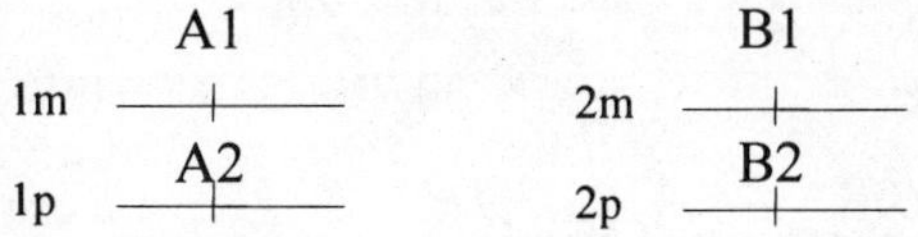

Gamete combinations:

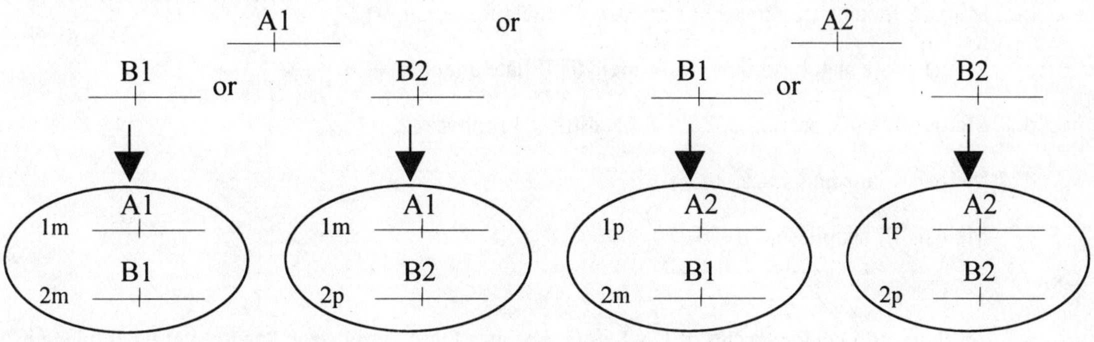

Q3-3. The sequence of genes found on one of the sister chromatids of a dyad is not necessarily identical to the sequence of genes found on a bivalent occurring in meiosis. Why? The answer: crossing over, which occurs early during prophase, results in the exchange of pieces of chromatids from homologous chromosomes. Crossing over takes place before the chromosomes have completely condensed and tetrads are visible.

Q3-4. Meiosis is a reduction division in which the number of chromosomes is reduced. The number of chromosomes is halved during the first meiotic division. The cells entering meiosis II contain one set of chromosomes and are therefore haploid.

Q3-5. Sperm cells would be expected to have more new mutations because spermatogonia are continuously dividing mitotically. Primary oocytes have already entered meiosis I before the female's birth. These cells are not continually dividing mitotically throughout the female's life.

Q3-6.

Feature	Mitosis	Meiosis
a	one	two (meiosis I and II)
b	two identical diploid (2N) cells	four usually dissimilar haploid (N) cells
c	once, during S phase	once, during S phase
d	NO	YES
e	NO	YES
f	YES	NO
g	growth, repair	maintenance of chromosome number during sexual reproduction; generation of variation

Answers to Even Numbered End of Chapter Exercises, Problems and Critical Thinking Questions

2. Kinetochore: chromosomal attachment point for spindle fibers, located on the centromere. Centromere: nucleoprotein constriction in eukaryotic chromosome on which kinetochore lies.

4. a. 46; b. 0; c. 23; d. 23; e. 23

6.

a. Mitosis, prophase, $2n = 6$, or meiosis II, prophase, $2n = 12$

b. Mitosis, metaphase, $2n = 6$, or meiosis II, metaphase, $2n = 12$

c. Mitosis, late anaphase, $2n = 6$, or meiosis II, late anaphase, $2n = 12$

d. Meiosis I, early prophase, $2n = 6$, or mitosis, prophase, $2n=3$

e. Meiosis I, late anaphase, $2n = 6$

f. Meiosis II, anaphase, $2n = 6$

8.

a. $2c$; there will be $2c$ amount of DNA in G_1 because DNA replication has not yet taken place in this diploid cell.

b. $4c$; there will be $4c$ amount of DNA in G_2 because DNA replication has taken place. $2c$; at the end of meiosis I, a cell will have $2c$ amount of DNA, disregarding differences brought about by differences in sex chromosomes. Since a cell entering meiosis I has $4c$ DNA, it will have $2c$ at the end of the first meiotic division and c at the end of the second meiotic division (one tetrad reduced to one chromosome).

10.

	$2n$ (= Dyads)	Bivalents (= Tetrads)
Human being	46	23
Garden pea	14	7
Fruit fly	8	4
House mouse	40	20
Roundworm	2	1
Pigeon	80	40
Boa constrictor	36	18
Cricket	22	11
Lily	24	12
Indian fern	1,260	630

12. The terms *reductional* and *equational* refer to the segregation of chromosomes during meiosis. The first division is reductional because homologous chromosomes are separated from each other, reducing the number of chromosomes by half. The second meiotic division is equational because the number of chromosomes remains the same although the number of chromatids is halved.

14. a.1/16 b. 1/16 c. 3/8. The number of chromosomal combinations is 2^n, where n = the number of different chromosomes. Thus, there are $2^4 = 16$ possible combinations. The chance of getting one member of a set is 1/2; therefore, the chance of getting all four from one parent is $(1/2)^4 = 1/16$. The possible combinations with two chromosomes from one parent are: ABC′D′, A′B′CD, A′BC′D, AB′CD′, A′BCD′, and AB′C′D (6/16 = 3/8).

16.

a. Since the diploid number is different for each species, the gametes will have twenty-five and nineteen chromosomes, respectively. Even if all nineteen of the red fox chromosomes were identical to nineteen of the arctic fox, some chromosomes could not pair, and imbalanced gametes and zygotes would result.

b. There must be some homology between the large chromosomes of the red fox and the small chromosomes of the arctic fox. This homology allows some of the chromosomes to pair.

18. Forty sperm, ten ova

20. a. 200 b. 100 c. 50. The primary spermatocyte is diploid and will undergo meiosis to yield 4 cells; $4 \times 50 = 200$. The secondary spermatocyte has completed one meiotic division and has one more division to go; $2 \times 50 = 100$. The spermatid is haploid, and each spermatid matures into a sperm cell.

22. a. 20; b. 20; c. 30; d. 10 per nucleus; e. 10

24. Human being: see following figure. Pea plant: with the obvious differences between corn and peas aside, figure 3.30 describes the pea life cycle. *Neurospora*: see figure 6.20.

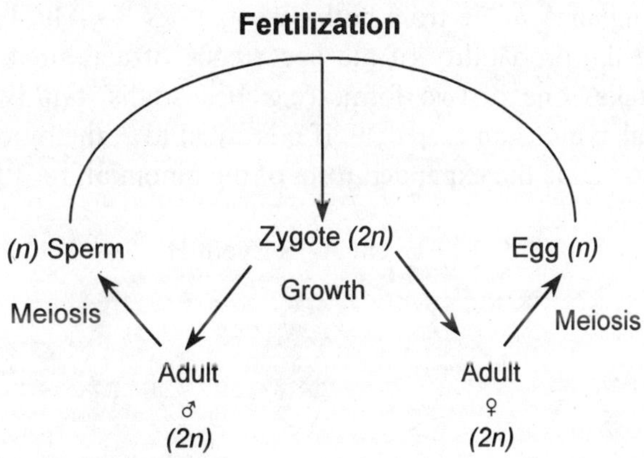

26. *AaBb* queen × *ab* drone yields

Sons (or gametes): *AB, Ab, aB, ab*

Daughters: *AaBb, Aabb, aaBb, aabb*

28. a. 21 b. 42 c. 63. In the pollen tube, the pollen nuclei result from meiosis and are therefore haploid. A leaf cell is diploid. The endosperm results from the fertilization by a haploid pollen nucleus of a diploid egg nucleus, yielding a triploid cell.

Critical Thinking Question

2. Presumably, the minimum number of chromosomes that a diploid organism could have is two; which is known for the roundworm, *Ascaris.* There is no theoretical upper limit, other than the physical limitation on the number of microtubules in a spindle. The fern, *Ophioglossum,* has 1,260 chromosomes, the highest number known. Recent research indicates that when chromosomal arms exceed half the spindle length in telophase, problems in fertility and viability follow. Also, chromosomes below 120,000-500,000 nucleotide base pairs are unstable. Clearly, evolution has selected for a reasonable number of chromosomes, closer to 50 than to 1,000.

CHAPTER 4
PROBABILITY AND STATISTICS

Key Genetic Concepts

Probability Theory: Probability is the expectation of the occurrence of a particular event. For example, the probability of a heterozygote (Aa) producing a gamete containing the A allele is ½. Individual probabilities can be combined to obtain the probability that two events will both occur. The probability that one of several mutually exclusive events will occur (unshaded areas of figure below) is equal to the sum of the individual event probabilities (**sum rule**). For example, the probability of producing a heterozygote in an F2 generation is equal to ¼ (probability of A from the female and a from the male) + ¼ (probability of A from the male and a from the female) or ½. The probability that two independent events will both occur (shaded area of figure below) is equal to the product of their individual probabilities (**product rule**). For example, the probability of obtaining an AA genotype in a F2 generation is equal to the probability of receiving an A allele from both parents (½ x ½). The **binomial theorem** can also be used to determine the probability of the occurrence of combinations of events where each individual outcome takes one of two forms (e.g. heads/tails, tall/dwarf, or A/a). The general formula for a binomial expression is $(p+q)^n$. If n is equal to 2, the binomial expression is $(p+q)^2$. Squaring the term p+q yields the expanded form of the binomial: $p^2 + 2pq + q^2$.

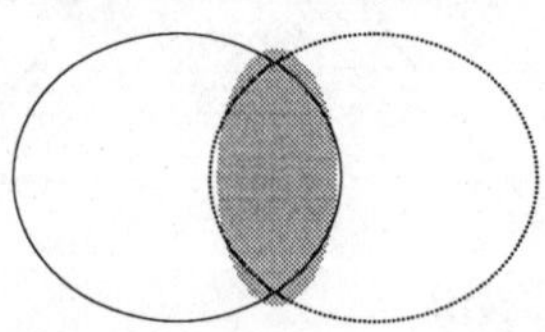

Statistics: Statistics are used by geneticists to design experiments, to summarize data, and to determine whether or not the results of an experiment support the hypothesis being tested. The **chi-square test** is a statistical test frequently employed by geneticists to determine whether or not a set of data conform to an established genetic ratio. For example, a geneticist may find that a testcross yields 50 individuals with the dominant phenotype and 40 with the recessive phenotype. The question is *Do these data conform to the 1:1 ratio hypothesized for a testcross (the* ***null hypothesis****)?* Geneticists perform a chi-square test and decide on the basis of the outcome of the test.

Problem-Solving Hints

Hint 1: The sum (*either/or*) rule is used when you want to know the probability of obtaining either A *or* B. For example, a heterozygote is either Aa or aA, but never both.

Hint 2: The product (*and*) rule is used when you want to know the probability of obtaining both A *and* B. For example, to be a heterozygote you must receive both A and a.

Hint 3: Pascal's triangle is useful in expanding binomials where n is large because it permits rapid determination of coefficients for each term in the expansion.

Hint 4: The **critical chi-square value**, the value to which a calculated chi-square value is compared, is obtained from a table. You must know the degrees of freedom and choose a level of significance to determine the critical value. The **degrees of freedom** represent a count of independent categories. For tests of Mendelian ratios, the degrees of freedom are equal to the number of categories minus one. The **level of significance** traditionally used is 0.05.

Hint 5: The null hypothesis is rejected if the calculated chi-square value exceeds the critical value obtained from the table. If the calculated value falls below the critical value, the researcher fails to reject the null hypothesis.

Practice Problems

Problem 4-1. The diploid chromosome number in humans is 46. What is the probability that a woman will produce an egg that contains only paternal chromosomes or only maternal chromosomes? Disregard crossing over for this problem.

Problem 4-2. Maple sugar urine disease is a recessive autosomal disorder in humans. Assume that two carriers of maple sugar urine disease marry and start a family. If they have four children, what is the chance that the children will have the following phenotypes, in the following order: normal, normal, affected, normal? What is the probability that any one of the four will be affected?

Problem 4-3. Assume that 5 babies are born at a hospital on the same day. Assuming a 50:50 sex ratio, what is the probability of no girls? Of all girls? Of two girls?

Problem 4-4. Expand the trinomial $(p+q+r)^2$ and relate the terms in the expansion to ABO blood genotypes.

Problem 4-5. A geneticist observed 46 green seedlings and 11 albino (without chlorophyll) seedlings among the progeny of a cross between two green tomato plants. What hypothesis should the geneticist test? Apply the chi-square test to the data and interpret the results.

Problem 4-6. Mendel crossed pea plants with round and yellow seeds to a plant with wrinkled and green seeds. The F1 generation was composed of plants producing round and yellow seeds. A cross of the F1 plants resulted in the 315 plants producing round and yellow seeds, 108 plants with round and green seeds, 101 plants with wrinkled and yellow seeds, and 32 plants producing wrinkled and green seeds. Mendel proposed that this ratio approximated a 9:3:3:1 ratio. Use the chi-square test to statistically test this hypothesis. Test the hypothesis that the data approximate a 1:1:1:1 ratio. Clearly state your conclusions.

General Questions

Question 4-1. From the vantage point of a scientist, what advantage does the use of the chi-square test offer as a tool in decision making?

Solutions to Practice Problems

P4-1. The probability that egg contains only maternal chromosomes is $(1/2)^{23}$. The probability that the egg contains only paternal chromosomes is the same. Using the sum rule, the probability that the egg contains all maternal *or* all paternal chromosomes is $(1/2)^{23} + (1/2)^{23}$ or 2.38 x 10-7.

P4-2. In this example, the probability of producing a normal child is 3/4 and the probability of producing an affected child is 1/4. The probability of having four children (sequence: normal, normal, affected, normal) is calculated as $(3/4)^2 \times (1/4)^1 \times (3/4)^1 = 0.1055$. The probability that three will be normal and one affected, without specifying a birth order, will be $4(3/4)^3 \times (1/4)^1$ or 0.4219.

P4-3. The binomial corresponding to 5 children is $(p+q)^5$. Expanded, we have:

$$p^5 + 5p^4q + 10p^3q^2 + 10p^2q^3 + 5pq^4 + q^5$$

where: p = probability of a girl = 0.5
q = probability of a boy = 0.5

The probability of no girls: q^5 or $(1/2)^5 = 0.0313$

The probability of all girls: p^5 or $(1/2)^5 = 0.0313$

The probability of two girls: $10p^2q^3$ or $10(1/2)^2(1/2)^3 = 0.313$

P4-4. $(p+q+r)^2 =$

p^2 +	2pq +	2pr +	q^2 +	2qr +	r^2
I^AI^A	I^AI^B	I^AI^O	I^BI^B	I^BI^O	I^OI^O

P4-5. This ratio appears to be a 3:1 (green : albino) ratio. The hypothesis to be tested is: *the ratio obtained by the geneticist approximated a 3 green to 1 albino ratio.* The chi-square test should be used. The chi-square value is calculated: $\Sigma[(\text{observed} - \text{expected})^2 / \text{expected}]$.

Phenotype	Observed	Expected number
Green seeds	46	¾ x 57 = 42.75
Albino seeds	11	¼ x 57 = 14.25
	57	

$\chi^2 = (46-42.75)^2/42.75 + (11-14.25)^2/14.25$

$\chi^2 = 0.247 + 0.741$

$\chi^2 = 0.988$ (1 df)

Critical value (2 – 1 = 1 df; *Hint 4*) = 3.841

Fail to reject hypothesis because 0.988 < 3.841 (*Hint 5*).

P4-6. Testing the hypothesis that the data approximate a 9:3:3:1 ratio:

Phenotype	Observed	Expected number
round, yellow	315	9/16 x 556 = 312.75
round, green	108	3/16 x 556 = 104.25
wrinkled, yellow	101	3/16 x 556 = 104.25
wrinkled, round	32	1/16 x 556 = 34.75
Total	556	

$\chi^2 = (315\text{-}312.75)^2/312.75 + (108\text{-}104.25)^2/104.25 + (101\text{-}104.25)^2/104.25 + (32\text{-}34.75)^2/34.75$

$\chi^2 = 0.016 + 0.135 + 0.101 + 0.218$
$\chi^2 = 0.47$ (3 df)

Fail to reject null hypothesis because $0.47 < 7.815$ (critical value for 3 df).

The data approximate a 9:3:3:1 ratio.

Testing the hypothesis that the data approximate a 1:1:1:1 ratio:

Phenotype	Observed	Expected number
round, yellow	315	1/4 x 556 = 139
round, green	108	1/4 x 556 = 139
wrinkled, yellow	101	1/4 x 556 = 139
wrinkled, round	32	1/4 x 556 = 139
Total	556	

$\chi^2 = (315\text{-}139)^2/1389 + (108\text{-}139)^2/139 + (101\text{-}139)^2/139 + (32\text{-}139)^2/139$
$\chi^2 = 222.849 + 6.914 + 10.389 + 82.367$
$\chi^2 = 322.519$ (3 df)

Reject null hypothesis because $322.519 > 7.815$ (critical value for 3 df).

The data *do not* approximate a 1:1:1:1 ratio.

Answers to General Questions

Q4-1. Statistics provide geneticists with a consistent decision making tool. Without statistics, a geneticist would need to rely on his/her "gut feeling" when interpreting data. While this means of examination may be effective in some cases, it does not provide a clear decision in all cases. For example, most people would agree that 49:51 closely approximates a 1:1 ratio, even without the benefit of statistics. However, the interpretation of a 60:40 ratio is substantially less clear. You are as likely to find geneticists who are willing to accept 60:40 as a 1:1 ratio as you are to find geneticists who do not accept this ratio. When statistics are used, the decision is more clear and a larger number of geneticists will agree on the interpretation.

Answers to Even Numbered End of Chapter Exercises, Problems and Critical Thinking Questions

2. Parents are heterozygotes ($Tt \times Tt$)

a. 1/4 = 0.25

b. 1/8 = 0.125 ([1/4][1/2])

c. $(6!/2!2!2!)(1/8)^2(1/8)^2(3/8)^2 = 0.0030899$

d. ~~3/8 = 0.375~~ 3/8 = 0.375

4.

a. 1/10,000 = 0.0001

b. $1/10{,}000 = 0.0001$

c. $(1/10{,}000)^2 = 0.00000001$

6. Assume that Mendel was observing the phenotypes of offspring produced from seeds obtained by testcrossing the plant having a dominant phenotype (*A-* × *aa*). As soon as an offspring with the recessive phenotype appeared ($p = 1/2$), he was certain that the dominant parent was heterozygous. If he tested five seeds and all produced plants with the dominant phenotype, there would be a 3% chance that the parent plant was still heterozygous ($p = [1/2]^5$ of getting only *A* alleles from an *Aa* genotype). Testing seven seeds would lower the chance of heterozygosity to less than 1%. "Pretty reliable" is arbitrary; complete certainty requires an infinite number of offspring.

8. a. 1/4 b. 6/16 c. 1/16. The chance of a recessive genotype (*aa*) from two heterozygotes (*Aa*) is 1/4. For an either-or situation, we add probabilities of being one or the other but not both, 3/16 + 3/16 = 6/16. For two independent events to occur simultaneously, we multiply individual probabilities, 1/4 × 1/4 = 1/16.

Note that parts b and c can be seen as outcomes of the 9:3:3:1 ratio.

10. a. 27/64 b. 3/512. With four children, the formula for unordered events is:

$$\frac{4!}{1!3!}(1/4)(3/4)^3 = 4(1/4)(27/64) = 27/64$$

The chance of a normal boy = (3/4 normal) (1/2 boy) = 3/8; affected boy or girl = (1/4)(1/2) = 1/8. We multiply these probabilities together to get 3/8 × 1/8 × 1/8 = 3/512.

12. a. 1/3 b. 5/9. We calculate the frequency of deaths from cancer as 300/900 = 1/3, and the deaths due to heart disease as 200/900 = 2/9. For death either by cancer or heart disease, we add the probabilities, 1/3 + 2/9 = 5/9.

14. $5(1/2)^5 = 5/32$. The cross is *Tt* × *tt*. This is a testcross, so there is a one-half chance of either a taster or nontaster offspring. The binomial expression is used for unordered events:

$$P = \frac{5!}{4!1!}(1/2)^4(1/2) = 5(1/2)^5$$

16. a.Hypothesis: *Rr* × *Rr* produces *R-*:*rr* in a 3:1 ratio. Chi-square, one degree of freedom (d.f.), = 0.263. Critical chi-square at 0.05, 1 d.f., = 3.841.

	Round	*Wrinkled*	Sum
Observed	5,474	1,850	7,324
Expected	3/4	1/4	
	5,493	1,831	7,324
$O - E$	−19	19	
$(O - E)^2$	361	361	
$(O - E)^2/E$	0.066	0.197	$0.263 = \chi^2$

Fail to reject (i.e., accept) the null hypothesis at 0.05 level and therefore also at the 0.01 level.

b. Hypothesis: *Rr* × *Rr* produces *R-*:*rr* in a 3:1 ratio. Chi-square, 1 d.f., = 0.474. Critical chi-square at 0.05, 1 d.f., = 3.841.

	Round	*Wrinkled*	Sum
Observed	45	12	57
Expected	3/4	1/4	
	42.75	14.25	57
$O-E$	2.25	–2.25	
$(O-E)^2$	5.063	5.063	
$(O-E)^2/E$	0.118	0.355	$0.474 = \chi^2$

Fail to reject (i.e., accept) the null hypothesis at 0.05 level and therefore also at the 0.01 level.

c. Hypothesis: *Rr*:*RR* are in a 2:1 ratio. Chi-square, 1 d.f., = 0.174. Critical chi-square at 0.05, 1 d.f., = 3.841.

	Rr	*RR*	Sum
Observed	372	193	565
Expected	2/3	1/3	
	376.67	188.33	565
$O-E$	–4.67	4.67	
$(O-E)^2$	21.81	21.81	
$(O-E)^2/E$	0.058	0.116	$0.174 = \chi^2$

Fail to reject (i.e., accept) the null hypothesis at 0.05 level and therefore also at the 0.01 level.

d. Hypothesis: *VvLl* × *vvll* produces a 1:1:1:1 ratio of offspring. Chi-square, 3 d.f., = 1.084. Critical chi-square at 0.05, 3 d.f., = 7.815.

	VvLl	*vvLl*	*Vvll*	*vvll*	Sum
Observed	47	40	38	41	166
Expected	1/4	1/4	1/4	1/4	
	41.5	41.5	41.5	41.5	166
$O-E$	5.5	–1.5	–3.5	–0.5	
$(O-E)^2$	30.25	2.25	12.25	0.25	
$(O-E)^2/E$	0.729	0.054	0.295	0.006	$1.084 = \chi^2$

Fail to reject (i.e., accept) the null hypothesis at 0.05 level and therefore also at the 0.01 level.

18. Hypothesis: $AaCc^a \times AaCc^a$ produces agouti (*A-C-*), black (*aaC-*), albino ($--c^a c^a$) offspring in a 9:3:4 ratio. Critical chi-square, two degrees of freedom at probability of 0.05, = 5.991.

	Agouti	*Black*	Albino	Total
Observed	28	7	13	48
Expected	9/16	3/16	4/16	
	27	9	12	48
$O-E$	1	−2	1	
$(O-E)^2$	1	4	1	
$(O-E)^2/E$	0.037	0.444	0.083	$0.565 = \chi^2$

Since this chi-square, 0.565, is less than the critical chi-square, we fail to reject our hypothesis of two-locus genetic control with epistasis.

20. a. 9:3:3:1 b. yes. We have two genes, and if they are not linked, we expect a 9:3:3:1. To test the results, set up a chi-square table.

	O	*E*	*(O – E)*	*(O – E)2*	*(O – E)2/E*
long, tan	84	90	−6	36	36/90 = 0.4
long, dark	27	30	−3	9	9/30 = 0.3
short, tan	35	30	5	25	25/30 = 0.833
short, dark	14	10	4	16	16/10 = 1.6

Chi-square = 0.4 + 0.3 + 0.83 + 1.6 = 3.133. With three degrees of freedom, $p > 0.05$ (critical chi-square = 7.815). The observed variation falls within the expected random variation given that the hypothesis is correct. In other words, we fail to reject the null hypothesis.

Critical Thinking Question

2. One-half of the families would stop at two children, one-half of the remaining families would stop at three children, and so on. Thus the average family size is the sum of (1/2)(2) + (1/4)(3) + (1/8)(4) + (1/16)(5) +..., or

Average family size =

$$\sum_{n=1}^{\infty} (1/2^n)(n+1)$$

$$= \sum_{n=1}^{\infty} (n+1)/2^n$$

This summation reaches a limit of 3.0 at infinity.

CHAPTER 5
SEX DETERMINATION, SEX LINKAGE, and PEDIGREE ANALYSIS

Key Genetic Concepts

Mechanisms of Sex Determination: Sex determination is usually a complex process involving a series of developmental changes under genetic and hormonal control. In humans, social factors also play a significant role in the formation of a complete sexual identity. Examples of sex determining mechanisms are discussed below.

- **Ploidy**: The number of complete sets of chromosomes an individual possesses determines sex in a group of insects which includes the bees, ants, and wasps (the Hymenoptera). In this case, females are diploid and result from fertilized eggs while males are haploid and arise from unfertilized eggs.
- **Environmental Cues**: In some species, sex is determined by the environmental cues (e.g. temperature) experienced during development.
- **Sex Chromosomes**: There are three basic types of sex-determining mechanisms involving genes on specific chromosomes: **XY, XO,** and **ZW**. In the XY system, females have a pair of morphologically identical chromosomes (a **homomorphic** pair, XX). Females produce only one type of gamete with respect to the sex chromosomes and are referred to as **homogametic**. Males have a pair of morphologically dissimilar chromosomes (a **heteromorphic** pair, XY) and therefore produce two kinds of gametes (**heterogametic**). Human beings and fruit flies have XY sex chromosomes. In contrast, female birds are heteromorphic and herterogametic (ZW) while males are homomorphic and homogametic (ZZ). Finally, in some species of insects, for example, there is only one type of sex chromosome. Females are usually XX while males are X (written XO to indicate that the normal complement is a single X chromosome).
- In *Drosophila*, genes favoring maleness are found on the autosomes while genes favoring femaleness are located on the X chromosome. The ratio of X chromosomes to sets of autosomes determines if the individual is male or female. Males result if the ratio is 0.5 (1 X chromosome / 2 sets of autosomes) and females result if the ratio is 1.0 (2 X chromosomes / 2 sets of autosomes). This mechanism of sex determination was described by Calvin Bridges as the **Genic Balance Hypothesis**.
- In human beings, the presence of a **sex-determining Y (SRY) gene** initiates the development of males. This gene is located on the Y chromosome. Several XX males have been identified. One of the X chromosomes of these males contains an SRY gene.

Dosage Compensation: Human males and females have the same number of autosomal chromosome pairs (22) although they differ in the number of X-chromosomes they possess. Remember that the X chromosome is relatively large while the Y chromosome has very few functioning genes. This creates a problem: females receive two copies of the genes on the X chromosome while males receive only one copy. How do females compensate for this dosage difference? In mammals, one of the female's X chromosomes is inactivated. Examination of the nucleus of female derived cells reveals the presence of a darkly staining body that is absent in cells obtained from normal males. This structure, a **Barr body**, is believed to be an inactivated X chromosome. The **Lyon hypothesis** suggests that only one X chromosome is active in a cell. The inactivation occurs early during embryonic development and is a random process in most cases. A small bank of genes at the tip of the X chromosome are required in double dose for normal female development. These genes are not inactivated. In fruit flies, dosage compensation is achieved by the hyperactivity of the male's single X chromosome.

Traits on the Sex Chromosomes: Genes located on the X or Y chromosome follow different patterns of inheritance than genes located on the autosomes. This difference arises from the fact that females possess two X chromosomes while males have only a single X. Likewise, only males possess a Y chromosome. The patterns of inheritance for each type of linkage related to sex are described below.

- **Holandric Traits** are those found only in males. The genes controlling these traits are found on the Y-chromosome. Males pass their Y chromosome to all of their sons and none of their daughters which accounts for the observed male to male pattern of inheritance.
- **Sex Linked Traits** also show a unique pattern of inheritance. Males inherit only a single copy of genes on the X chromosome from their mothers and are described as **hemizygous**. Since males inherit only one allele, that allele is expressed. Males pass their single X chromosome to all of their daughters but none of their sons. Conversely, females inherit two copies of sex linked traits: one allele from the mother and one allele from the father. Crosses involving sex linkage can be easily identified: the phenotypes of males and females will differ in one or more generations. In addition, switching the genotype of the male and female parent (constructing a **reciprocal cross**) will yield different results.

Traits Related to Sex: Several traits are related to sex although the genes are not necessarily located on the sex chromosomes. The genes for some traits are present in both sexes while the trait is expressed in only one. These traits, such as breasts in females, are said to be **sex-limited**. **Sex-influenced** traits, such as pattern baldness, appear in both sexes but are expressed more frequently in one sex.

Pedigree Analysis: Pedigrees, or family trees, represent the ancestry of a group of related individuals. Family trees are useful for examining a pattern of inheritance in a genetic study.

Relationship Between Genotype and Phenotype: In many human traits, the genotype of an individual does not directly determine the phenotype expressed. In traits with reduced **penetrance**, the genotype may not be expressed in the phenotype. This results in an "all-or-nothing" expression pattern. In other cases, the genotype may be expressed in varying degrees, from mild to extremely severe. This variable expression of a genetic trait is referred to as **expressivity**. Finally, there are examples of human traits (**phenocopies**) that appear to be genetically controlled when they are the result of environmental influences.

Problem-Solving Hints

Hint 1: When solving sex linkage problems, symbols are placed on the X (or Z) chromosome but not on the Y (or W) chromosome. Place symbols for only sex linked traits on the X-chromosome. Symbols for autosomal genes should be used as previously described in Chapter 2.

Hint 2: Sex linkage is *suggested* if the phenotypes of males and females differ. Sex linkage is also suggested if the results of reciprocal crosses yield different progeny phenotypes.

Hint 3: Sex linked traits often show a **criss-cross pattern of inheritance**. Females pass an X chromosome to all of their offspring. Males pass their X-chromosome to their daughters and their Y-chromosome to their sons. Unique phenotypic ratios result because males are hemizygous and females are not.

Hint 4: Sex linked dominant inheritance is *suggested* if all of the daughters but none of the sons of an affected male are affected. Dominant inheritance is also suggested in an affected female produces at least fifty percent affected children.

Hint 5: Sex linked recessive inheritance is *suggested* if most affected individuals are males. Affected females will produce only affected sons. Carrier females will produce a 1:1 ratio of affected : normal sons. Affected females result from carrier (or affected mothers) *and* affected fathers.

Hint 6: In mammals, dosage compensation alters the expression of the genotype in heterozygous females. The appearance of mosaic females may *suggest* sex linkage.

Hint 7: In the F2 generation of a cross, a 3:3:1:1 phenotypic ratio *suggests* one sex-linked trait and one autosomal trait.

Practice Problems

Problem 5-1. In *Drosophila melanogaster*, the allele for miniature body is recessive and located on the X-chromosome. Design a cross of true-breeding flies which will allow the observer to distinguish males from females based on the size of the individual.

Problem 5-2. In the laboratory, you cross a homozygous female fruit fly with bobbed bristles and warty eyes to a homozygous wild-type male. Assuming these two traits are recessive and sex linked, predict the genotypes and phenotypes of the progeny produced in the F1 and F2 generations. How would the results have differed if the traits under study were sex linked dominant? Predict the results of the cross assuming that the traits are autosomal recessive.

Problem 5-3. Assume that you are a geneticist studying a new, rare genetic disorder. You are presented with a man (A), who is affected, and his normal wife. The couple has five children: two affected daughters and three normal sons. Each of the sons is married to a normal female and all have normal children. One of the affected daughters married a normal man and had two sons: one with the disorder and one without the disorder. From this information, determine the genotype of each individual represented. What can you conclude about the parents of male A?

Problem 5-4. Lisa and Tina are identical (monozygotic) twins. However, Lisa is colorblind while Tina has normal color vision. Explain this difference on the basis of the Lyon hypothesis.

Problem 5-5. For each of the sex chromosome complements below, determine the sex of each individual and the number of Barr bodies present in humans. For fruit flies, provide the sex of each individual if two sets of autosomes are present and if three sets of autosomes are present.

Chromosome Complements: X, XX, XY, XXX, XXY, XYY

Problem 5-6. Consider the plant *Lychnis alba* which has an XY sex determining mechanism. If broad-leaved females are crossed to narrow-leaved males, the F1 generation is composed of all broad-leaved individuals. In the F2, all females are broad-leaved and males are either broad-leaved or narrow-leaved. Which sex is homogametic and which is heterogametic?

Problem 5-7. Horns in sheep are governed by a sex-influenced gene. This trait is dominant in males but recessive in females. Provide all genotypes and phenotypes possible in male sheep and in female sheep.

Problem 5-8. Use the diagrams below to illustrate reduced penetrance and expressivity. Assume that each figure represents an individual female or male with the same genotype. For individuals that are unaffected, leave the circle blank. Use varying degrees of shading to illustrate (a) reduced penetrance and (b) variable expressivity.

(a)

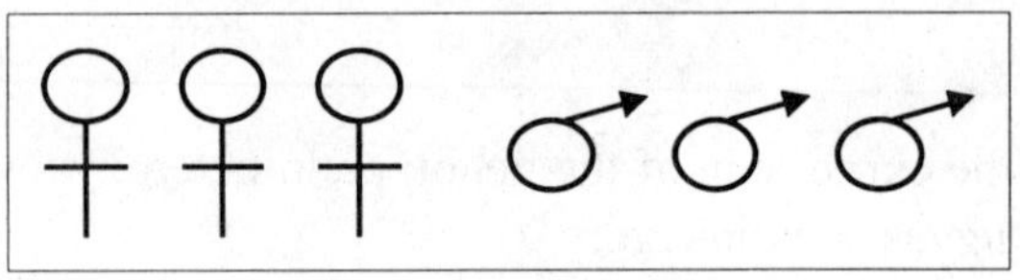

(b)

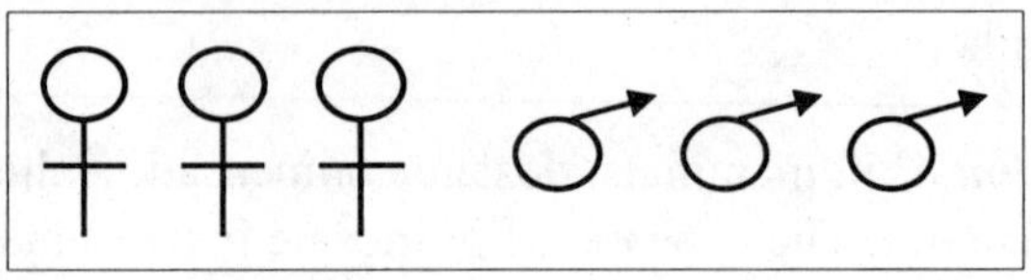

Problems 5-9. Complete the pedigree below such that (a) Y-linked inheritance is suggested; (b) recessive autosomal inheritance in suggested; (c) recessive sex-linked inheritance is suggested.

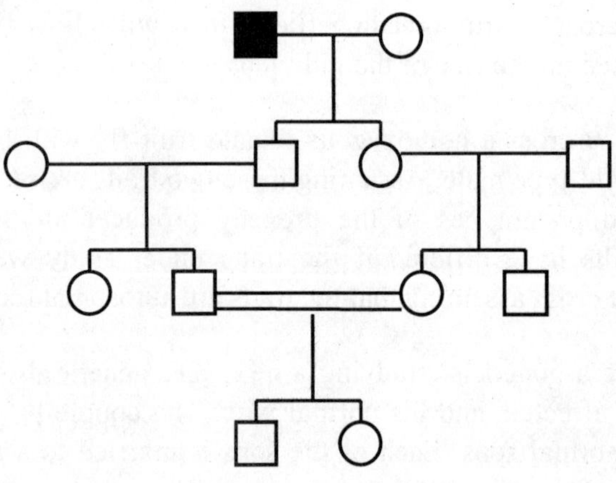

Solutions to Practice Problems

P5-1. There are two possible crosses which can be designed. In the first, a miniature male is crossed with a wildtype female. In the second, reciprocal, cross a wildtype male is crossed with a miniature female. As is obvious from the diagrammed crosses below, number two yields F1 males and females with different phenotypes (*Hint 2b*)

X^+ = chromosome bearing wild-type allele X^m = chromosome with allele for miniature body (*Hint 1*)

Cross 1:

P1: X^mY x X^+X^+

Female gametes

Male gametes	X^+	X^+
X^m	X^+X^m	X^+X^m
Y	X^+Y	X^+Y

F1: Normal females and males

Cross 2:

P1: X^+Y x X^mX^m

Female gametes

Male gametes	X^m	X^m
X^+	X^+X^m	X^+X^m
Y	X^mY	X^mY

F1: Normal females and miniature males

P5-2. B = allele for normal bristles, b = allele for bobbed bristles, W = allele for normal eyes, w = allele for warty eyes

Sex linked recessive inheritance:

P1: $X^{bw}X^{bw}$ x $X^{BW}Y$

F1: $X^{BW}X^{bw}$ x $X^{bw}Y$ wild-type females x bobbed, warty males

F2:

	X^{BW}	X^{bw}
X^{bw}	$X^{BW}X^{bw}$	$X^{bw}X^{bw}$
Y	$X^{BW}Y$	$X^{bw}Y$

½ wild-type, ½ bobbed, warty

Sex linked dominant inheritance:

P1: $X^{BW}X^{BW}$ x $X^{bw}Y$

F1: $X^{BW}X^{bw}$ x $X^{BW}Y$ all bobbed, warty

F2: $X^{BW}X^{BW}$ $X^{BW}X^{bw}$ bobbed, warty females

$X^{BW}Y$ $X^{bw}Y$ ½ bobbed, warty males and ½ wild-type males

Autosomal recessive inheritance: (refer to Chapter 2)

P1: BBWW x bbww	
F1: BbWw x BbWw	all individuals wild-type
F2: 9:3:3:1 phenotypic ratio	9/16 wildtype
	3/16 wild-type bristles, warty eyes
	3/16 bobbed bristles, wild-type eyes
	1/16 bobbed, warty

Note: No differences between males and females in either generation.

P5-3. The first clue to the solution of this problem is the distribution of affected individuals. Note that a criss-cross pattern of inheritance is observed: affected male to affected daughters (*Hint 3*). This indicates that the trait is sex-linked. The fact that females are affected indicates that the trait is dominant (*Hint 4*). We can conclude that the man's mother was affected. We cannot determine the genotype of his father. A = allele for disorder a = normal allele

Mother	Father	Man	His Wife	His Sons	Daughter
$X^AX^?$	$X^?Y$	X^AY	X^aX^a	X^aY	X^AX^a

P5-4. Colorblindness is an X-linked trait in human beings. While identical twins share all of their genes in common, differences may arise in sex linked traits because of X-inactivation (*Hint 6*). Assume that both women are heterozygotes. Lisa could be colorblind if, by random chance, the X chromosome bearing the colorblind allele is active in the cells of her retina. The X chromosome possessing the normal allele is inactive in these cells. The opposite pattern would be observed in Tina: the X chromosome bearing the normal color vision allele is active.

P5-5. Recall that sex in *Drosophila* is determined by the ratio of X chromosomes to sets of autosomes. In human beings, sex is determined by the presence of the SRY gene on the Y chromosome. Note also that Barr bodies are found in human beings with two or more X chromosomes regardless of their sex. Barr bodies are inactivated X chromosomes, no dosage compensation occurs with the Y chromosome. Bar bodies are not found in fruit flies.

Sex determination and dosage compensation in human beings.

Sex Chromosomes	Sex of individual	Number of Barr bodies present
X	Female	0
XX	Female	1
XY	Male	0
XXX	Female	2
XXY	Male	1
XYY	Male	0

Sex determination in the fruit fly, *Drosophila melanogaster*.

Sex Chromosomes	Sex if individual has **two** sets of autosomes	Sex if individual has **three** sets of autosomes
X	ratio 0.5 = male	ratio 0.33 = metamale
XX	ratio 1.0 = female	ratio 0.67 = intersex
XY	ratio 0.5 = male	ratio 0.33 = metamale
XXX	ratio 1.5 = metafemale	ratio 1.0 = female
XXY	ratio 1.0 = female	ratio 0.67 = intersex
XYY	ratio 0.5 = male	ratio 0.33 = metamale

P5-6. Examination of the F1 reveals that all individuals are broad-leaved which indicates that this trait is dominant. Notice that the females in the P1 generation possess the dominant trait. If females are the *heterogametic* sex, the female will pass an X chromosome bearing a dominant allele to all of her sons, but none of her daughters. This would result in an F1 generation composed of broad-leaved males and narrow-leaved females. This is not the case. If the female is the *homogametic* sex, she will pass an X chromosome bearing a dominant allele to all of her offspring. In the F1, both sexes will express the dominant phenotype. The observed F1 phenotypes are consistent with the later hypothesis (females are homogametic).

P5-7. This problem provides you with the solution to the problem: horn presence is sex-influenced. Recall that both males and females possess the alleles for a sex-influenced trait although the trait occurs more frequently in one sex or the other. The presence of horns is dominant in males but recessive in females. The alleles are not located on the X chromosome. Genotypes and phenotypes are outlined below.

Genotype	Male Phenotype	Female Phenotype
HH	Horned	Hornless
Hh	Horned	Hornless
hh	Hornless	Horned

P5-8.

(a) (b)

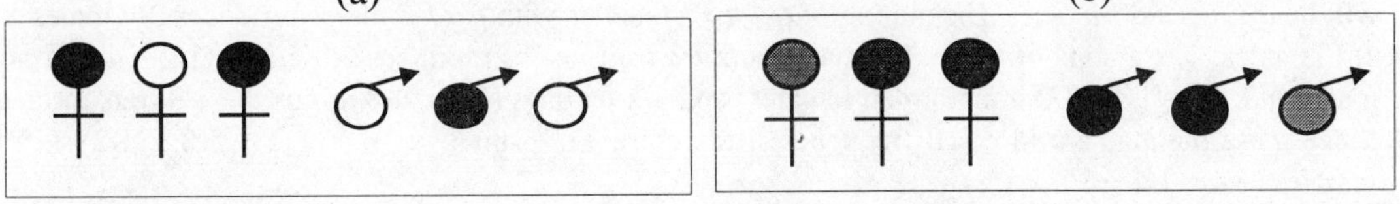

P5-9. Notice that a Y-linked trait is passed from male to male only. No females are affected or carry the allele. The recessive autosomal trait depicted in (b) is expressed only when two carriers produce children. Sex-linked recessive inheritance shows a criss-cross pattern of inheritance (*Hint 5*). Likely carriers are noted.

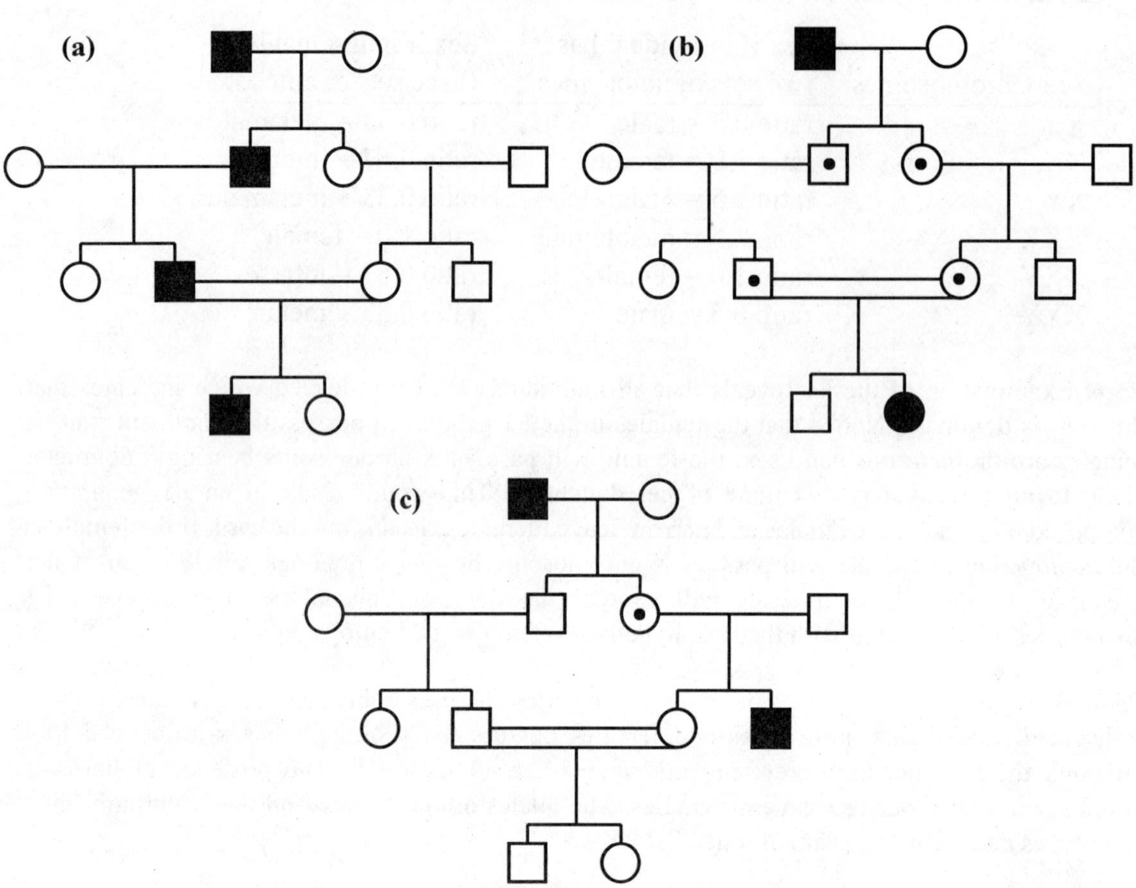

Answers to Even Numbered End of Chapter Exercises, Problems and Critical Thinking Questions

2. Apparent ratio of offspring is 3:1, male to female. Fertile males are 50% *tra tra* and 50% tra^+*tra*. Fertile daughters are all tra^+ *tra*; daughters counted as sterile males are *tra tra* = 1/4 of total. If these progeny are mated among themselves, all females will be *tra* tra^+ (*tra tra* females are sterile). Males will be *tra tra* and *tra* tra^+. The mating of *tra* tra^+ females with *tra tra* males produces 3/4 males and 1/4 females. The mating of *tra* tra^+ females with *tra* tra^+ males produces 5/8 males (1/2 = 4/8 normal males and 1/8 = 1/4 × 1/2 transformed females) and 3/8 females (1/2 = 4/8 minus the 1/8 transformed). Since these matings are in equal frequencies, there are 11/16 males (3/4 × 1/2 + 5/8 × 1/2) + 5/16 females (1/4 + 1/2 + 3/8 × 1/2).

4. Sex switches are genes that determine the developmental pathway of an organism, either toward male or female, depending on the presence or state of the gene. *SRY* is the male sex switch in human beings, whereas *Sxl* is a female switch in *Drosophila.*

6. The pattern should be fifteen bands made up of combinations of A, B_1, and B_2 subunits combined four at a time. These are both allozymes of B_1 and B_2 subunits and isozymes of A and B subunit interactions.

8. Early onset in calico cats; tortoiseshell cats have a later onset of Lyonization as seen by the fact that there were more cells present at that time and thus smaller sectors.

10.

	Cross	*Reciprocal*
P_1 female	Z^pW	Z^+W
male	Z^+Z^+	Z^pZ^p
F_1 female	Z^+W	Z^pW
male	Z^+Z^p	Z^+Z^p
F_2 females	Z^+W, Z^pW	Z^+W, Z^pW
males	Z^+Z^+, Z^+Z^p	Z^+Z^p, Z^pZ^p

12. Pseudodominance refers to the hemizygous (one copy) appearance of a recessive trait. A phenocopy is an environmentally induced phenotype similar to a genetically controlled one.

14. Color is sex-linked, black eye is dominant, males carry two alleles for color, and females carry one. Only the third cross alerts us to sex linkage, as we see a difference between males and females. The second cross indicates that black is dominant to pink. Recall that in birds, the male is the homogenetic (ZZ) sex. If we let B = black and b = pink, the crosses can be diagrammed as follows:

1. $Z^bZ^b \times Z^bW$		2. $Z^BZ^B \times Z^bW$		3. $Z^bZ^b \times Z^BW$	
↓		↓		↓	
Z^bZ^b	Z^bW	Z^BZ^b	Z^BW	Z^BZ^b	Z^bW
all pink - eyed		all black - eyed		black - eyed male	pinked - eyed female

16. a. Brown teeth is probably an X-linked dominant trait. b. 1/2. We see a difference in the phenotype of male and female children, suggesting some mode of sex linkage. Brown teeth cannot be Y-linked, or all the sons would have brown teeth. Since all the daughters and one of the sons have brown teeth, brown must be an X-linked dominant trait, and white is recessive. If B = brown teeth and b = white teeth, the original mating is $X^bX^b \times X^BY$. All daughters from this mating must be heterozygous. The A-B mating is therefore $X^BX^b \times X^bY$. One half of all children will be expected to receive the X^B chromosome.

18. Two genes are involved, one autosomal recessive that produces pink eyes, and one X-linked recessive that produces white eyes. Any fly homozygous or hemizygous for white eyes will have white eyes. The fact that the F_1 flies are wild-type indicates two different genes, and the different F_2 results for males and females indicate at least one X-linked gene. One half of the F_2 males have white eyes, a ratio expected for an X-linked recessive gene; we see 3:1 ratio for red:pink, a ratio expected for an autosomal gene. The F_2 females show a 3:1 ratio for red:pink, suggesting that all females have at least one normal X chromosome. If X^+ -A- = red, X^+ -aa = pink, X^w -A- = white, and X^wX^w- = white, the cross can be diagrammed as follows:

$$X^+X^+aa \quad \times \quad X^WYAA$$

$$\downarrow$$

$$X^+X^WAa \quad \times \quad X^+YAa$$

all red

$$\downarrow$$

$3X^+\text{-}A\text{-}$	red	$3X^+YA\text{-}$	red
$1X^+\text{-}aa$	pink	$1X^+Yaa$	pink
		$3X^WYA\text{-}$	white
		$1X^WYaa$	white

20. Abnormal-1 flies are homozygous for an autosomal recessive gene. Abnormal-2 flies are heterozygous for a dominant gene; it could be either autosomal or X-linked. Abnormal-3 flies are homozygous for a dominant gene that is either autosomal or X-linked. In the first case, the lack of difference in the sexes indicates autosomal inheritance. Since all females are normal, abnormal must be recessive. For abnormal-2, the 1:1 ratio within each sex indicates a heterozygote × homozygote mating: The cross could be $X^AX^a \times X^aY$ or $Aa \times aa$. For abnormal-3, the trait must be dominant, and we can't tell if it is X-linked or autosomal.

22. We see four phenotypes, so we must have at least two genes involved. We see no difference in body color between sexes, so we can conclude that body color is autosomally controlled. Note that if body color were X linked, males should have had dark bodies. Since all F_1 offspring have tan bodies, tan body must be dominant to dark body.

We see a difference in eye color between the sexes in the F_1 generation, suggesting that eye color is X linked. Since the F_1 females have red eyes, red must be dominant, and hence the original white-eyed female must be homozygous for the recessive allele.

In the F_2 generation, the ratio of tan to dark is 51:16—very close to the 3:1 ratio expected for one autosomal gene. Red to white is 36:31, very close to the expected 1:1 ratio for an X-linked trait.

Let dk^+ = tan body, dk = dark body, X^+ = red eyes, X^W = white eyes. Parents are $dkdk\ X^WX^W \times dk^+dk^+\ X^+Y$. We can then diagram the cross between the resulting F_1 individuals: $dk^+dk\ X^+X^W \times dk^+dk\ X^WY$. This cross yields

$$(1/2)X^+ \times (1/2)X^W \text{ (or Y)} \times 3/4 \text{ tan} = 3/16, \text{ male or female, red-eyed, tan-bodied}$$

$$\times 1/4 \text{ dark} = 1/16, \text{ male or female, red-eyed, dark-bodied}$$

$$(1/2)X^W \times (1/2)X^W \text{ (or Y)} \times 3/4 \text{ tan} = 3/16, \text{ male or female, white-eyed, tan-bodied}$$

$$\times 1/4 \text{ dark} = 1/16, \text{ male or female, white-eyed, dark-bodied}$$

24.

a. Could be an autosomal recessive, but most likely a sex-linked recessive, mode of inheritance. If complete penetrance is assumed, it could not be a dominant or Y-linked mode of inheritance.

b. Could be an autosomal recessive, but most likely an autosomal dominant, mode of inheritance. X- and Y-linked inheritance are ruled out.

c. Could be an autosomal recessive, or an X-linked recessive, mode of inheritance (if the allele causing the trait was brought into the pedigree by the affected's maternal grandmother). Other modes of inheritance are ruled out.

26. Pedigree (*a*) could represent either Y linkage or autosomal inheritance. Pedigree (*b*) represents an autosomal recessive pattern. Reduced penetrance simply means that any of the affected in these pedigrees could have a normal phenotype.

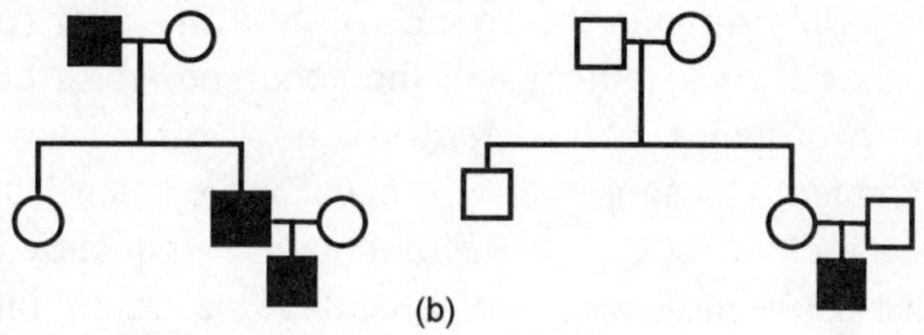

28.

a. autosomal recessive, or, most likely, X-linked recessive

b. X-linked dominant (most likely), autosomal dominant, autosomal recessive (unlikely)

c. Y-linked or autosomal dominant; unlikely: autosomal or X-linked recessive

In (a), the trait appears from normal individuals, indicating that it is recessive. In all generations, one-half of sons of normal women have the trait. It could also be autosomal, with parents of affected individuals being heterozygous. In (b), we notice that half of the sons of affected women have the trait, as well as all the daughters of affected men. The trait cannot be an X-linked recessive trait, for all sons of affected women would be affected. An autosomal dominant trait, heterozygous in affected individuals, would yield half the progeny affected. If the trait were autosomal recessive, each normal individual who marries an affected individual must be heterozygous, a rare possibility. In (c), the trait is passed only from father to son, behavior typical of a Y-linked trait. However, other modes are possible.

Critical Thinking Question

2. Assuming that all other methods of validation have failed, such as eye-witness accounts or fingerprints, then genetic methods can be used to see if the woman in question has alleles in common with the other members of the Tsar's family. Although we are familiar with electrophoresis of proteins, more sophisticated methods are known, called DNA fingerprinting, which we discuss in chapter 12. The woman's genetic profile can be compared either with the profile obtained from remains of the Tsar's family or by samples of direct descendents. This in fact did happen and genetic fingerprinting ruled that the woman was not the Princess.

CHAPTER 6
LINKAGE AND MAPPING IN EUKARYOTES

Key Genetic Concepts

Linkage: Each eukaryotic chromosome is a linear sequence of genes. Genes carried on the same chromosome are physically linked together. Therefore, genes within a **linkage group** are inherited together unless they are separated by crossing over. Consider two linked genes, A and B, each with two alleles. Assume that an individual with genotype AABB is crossed to an individual of genotype aabb to produce a dihybrid (AaBb). In the absence of crossing over, this dihybrid will produce only gametes with the allele combination found in the original parents (AB and ab, in our example). Only through the process of **crossing over** (physical exchange between chromatids of homologous chromosomes) will the genotype Ab or aB appear in a gamete. The distance between the two genes will determine how often new combinations of alleles (**recombination**) are formed: crossing over will occur more often if the genes are far apart and less frequently if they are close together. Recombination frequency can therefore be used as a measure of the distance between genes. One percentage of recombination is equal to one **map unit** or **centimorgan**.

Mapping Studies: A testcross of a heterozygous individual is used to determine the arrangement of genes on a chromosome and the distance between linked genes. Remember that a testcross of a dihybrid yields four phenotypic classes in a 1:1:1:1 ratio if the genes are located on separate chromosomes. In contrast, a testcross involving *linked* genes results in an excess of parental phenotypes (>50%) and a low number of recombinant types (< 50%). Offspring that have the same allele arrangement as the original parents are referred to as **parental** or **nonrecombinants**. Offspring with new allele arrangements resulting from crossing over are referred to as **recombinants** or **nonparental**. Because the homozygous individual contributes only recessive alleles to the offspring, the alleles received from the heterozygous parent are the ones expressed in the offspring. This allows parental and recombinant classes to be easily identified. The frequency of recombinant classes determines the distance between genes as mentioned above. Testcrosses involving two genes are referred to as a **two-point crosses**. Crosses involving three genes is referred to as a **three-point crosses**. When three linked genes are considered, single and **double crossovers** (those occurring between genes 1 and 2 and between 2 and 3) may be observed. **Crossover interference** may result if there are more or fewer double crossovers than expected. Specific details on how to solve two-point and three-point linkage map problems are given in your textbook.

Arrangement of Alleles: Heterozygous individuals receive different alleles from the maternal and paternal parent. When genes are linked, there are two possible arrangements of alleles in the heterozygote. First, the heterozygote can receive a chromosome bearing two dominant alleles (AB) from one parent and a chromosome containing two recessive alleles (ab) from the other. In this case, the allele configuration in the heterozygote is described as **cis** or **coupling**. If one of the chromosomes of the heterozygote contains the alleles Ab and the other chromosome bears the alleles aB, the configuration is described as **trans** or **repulsion**.

Haploid Mapping or Tetrad Analysis: This technique focuses on the total products of meiosis (**ascospores** retained in a sac called an **ascus**) in *Ascomycete* fungi. Meiosis of a diploid yeast cell results in the production of three ascus types: those with spores identical to the parental spores (**parental ditype**), those that are recombinants (nonparental **ditype**), and those that contain both classes of spores (**tetratype**). The loci under consideration are linked if the parental ditypes occur more frequently than nonparental ditypes.

Ordered spores in a bread mold such as *Neurospora*, can also be used to determine the distance between genes. In *Neurospora*, meiosis produces four haploid spores which each divide mitotically resulting in an ascus containing eight spores. The spores are ordered according to their centromeres which are denoted by the parental mating type from which they originated (for example A or a). In the absence of crossing over between the locus and its centromere, a 4:4 pattern (AAAAaaaa or aaaaAAAA) is observed. This pattern is referred to as a **first-division segregation** because segregation of the dyads during the first meiotic division determines the outcome. Crossing over results in the production of **second-division segregation** patterns (those resulting from segregations occurring during the second meiotic division) including 2:4:2 and 2:2:2:2. The number of first- and second-division segregates can be used to determine the distance between the gene and its centromere and to determine the order of genes that are linked to the same centromere.

Human Chromosome Mapping: Several methods can be used to map human genes to specific chromosomes. Pedigrees can be used effectively to map genes located on the X chromosome. For genes located on the autosomes, pedigrees can be used to determine the relationship between genes (e.g. do they occur together). **Somatic-cell hybridization** is used to localize genes to specific chromosomes. In this method, human and rodent cells are hybridized. The resulting cells are then tested for the presence or absence of specific phenotypes. This information is related to the presence or absence of specific human chromosomes.

Problem-Solving Hints

Hint 1: Although linkage changes the frequency of each phenotype, the number of phenotypic classes almost always remains the same. Statistical deviations from a 1:1:1:1 ratio (for a dihybrid testcross) or a 1:1:1:1:1:1:1:1 ratio (for a trihybrid testcross) indicate linkage.

Hint 2: In terms of numbers, the parental phenotypic classes will be the most frequent and the double crossover phenotypic classes will be the least frequent.

Hint 3: The double crossover phenotypic classes are used to determine the middle gene of a three gene sequence. In a double crossover, only the association of the middle gene differs when compared to the parental arrangement. For example, if the parental associations are ABC and abc, the double crossover types will have the associations AbC and aBc. Determination of the middle gene is essential. The placement of the other two genes is not critical (e.g. assuming the order is CBA rather than ABC will have no effect on the calculation of map distances).

Hint 4: Single crossover phenotypic classes cannot be reliably identified until the proper gene sequence has been established. Compare each single crossover class to the parental classes to determine where the crossover took place.

Hint 5: When solving three point linkage map problems, it is a good idea to use the gene symbols associated with each trait to represent the phenotypic description of testcross classes. Remember to use one symbol type (e.g. b or b+) for each gene under consideration.

Hint 6: In denoting linked genes, write the genotypes with a slash mark to separate maternally and paternally derived alleles (e.g. AB/ab rather than AaBb).

Hint 7: Taken together, the two parental class phenotypes represent the allele arrangement of the F1 female that was testcrossed. For example, assume that one of the parental classes obtained from the dihybrid testcross is mutant for both traits while the other parental class is wild-type for both traits. In this case, the alleles of the F1 dihybrid female were in the *cis* arrangement.

Hint 8: In some cases, three traits may be studied; however, only two genes are linked while the third is on a separate chromosome and assorts independently. This situation can be identified by examining the frequencies of each phenotypic class: four equally frequent majority classes and four equally frequent minority classes will be found. If the independently assorting gene is ignored, the problem reduces to a two-point linkage map.

Hint 9: Crossover interference reduces the likelihood of a double crossover occurring. If crossover interference is complete (I = 1), then no double crossovers are observed.

Practice Problems

Problem 6-1. In the tomato, stems are either purple (Aw) or green (aw) and the leaves are either wooly (Wo) or smooth (wo). Given the results of a dihybrid testcross (below), determine if the genes are linked. If so, how far apart are they on the chromosome? What is the linkage arrangement and the distance between genes? How would you describe the allele arrangement in the F1 heterozygote used in the testcross?

purple, wooly	205	green, wooly	810
purple, smooth	790	green, smooth	195

Problem 6-2. In *Drosophila*, a kidney shaped eye (k), cardinal colored eye (c), and ebony body (e) are recessive traits located on chromosome number three. A pure breeding wild-type male is crossed with a pure breeding female with kidney shaped, cardinal colored eyes and an ebony body. All of the F1's were wild-type. From the F1 testcross data given below, determine if the traits are linked. If so, calculate all map distances.

kidney, cardinal, ebony	1758	kidney	98
wild-type (all traits)	1768	ebony, cardinal	88
kidney, ebony	128	kidney, cardinal	12
cardinal	134	ebony	14

Problem 6-3. Consider three characteristics in the hypothetical animal, the zeebog from Madagascar. In the population, Madagascar zeebogs are either hairy (*H*) or hairless (*h*); crazy (*C*) or sane (*c*); and mischievous (*M*) or innocent (*m*). The results obtained from a testcross of a trihybrid female are given below.

hairy, sane, mischievous	195	hairy, crazy, innocent	790
hairy, sane, innocent	205	hairy, crazy, mischievous	816
hairless, sane, mischievous	782	hairless, crazy, mischievous	197
hairless, sane, innocent	815	hairless, crazy, innocent	200

What is the linkage arrangement and the distance between genes? How would you describe the allele arrangement in the F1 heterozygote used in the testcross?

Problem 6-4. Consider three linked genes (x, y, and z) on a single chromosome. Assume that a three-point linkage map showed that genes x and z are 30 map units apart. The x-y distance is 14 map units and the y-z distance is 16 map units. Devise a data set (N= 1000 individuals) with no crossover interference that supports these gene distances.

Problem 6-5. Considering the map distances given in problem 4, devise a data set (N= 1000 individuals) that supports the given gene distances assuming 100% crossover interference.

Problem 6-6. Considering again the map distances given in problem 4, devise a data set (N= 1000 individuals) that supports the given gene distances if crossover interference is 55%.

Problem 6-7. *Sordaria* is a mold that naturally grows on the dung of herbivores. It is an ascomycete and a close relative of *Neurospora.* In this species, spores may be black (wild-type) or tan in color. The distance between the mutant tan gene and the centromere has been reported as 26 map units. Assume that you cross a tan mutant strain to a wild-type strain and you score a total of 2,000. What data would you record if the above map distance is accurate?

Spore Sequence	**Frequency**
4:4	________
2:2:2:2	________
2:4:2	________

Problem 6-8. Somatic cell hybridization was used to map each of four enzymes to the human chromosome on which the loci reside. The results of the analysis are given below. (a) Which chromosome contains human enzyme 1a? (b) human enzyme 2a? (c) human enzyme 3a? (d) human enzyme 4a?

	Enzyme Activity				Human Chromosome Presence				
Clone	1a	2a	3a	4a	1	2	3	4	5
A	+	--	+	--	+	--	--	+	--
B	+	--	--	+	--	+	+	+	--
C	+	+	--	+	--	--	+	+	+
D	+	+	--	--	--	--	--	+	+
E	+	--	--	--	--	--	--	+	--

Problem 6-9. Examine the pedigree below. The traits under consideration are both X-linked and recessive. Note that the male in generation 1 expresses trait A. You should assume that the female in generation I carries the allele for trait B. From the information given, estimate the distance between the two genes. In what configuration (cis or trans) are the alleles of the female indicated with an arrow?

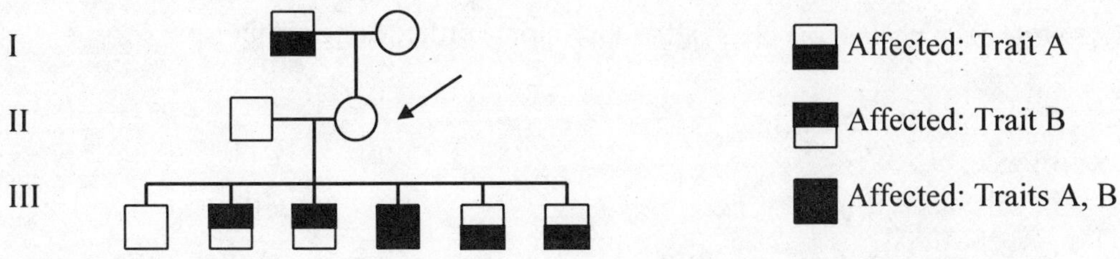

General Questions

Question 6-1. Consider a two-point mapping problem. Can crossing-over always be detected?

Question 6-2. Do linked genes always fail to assort independently during meiosis?

Solutions to Practice Problems

P6-1. The traits are linked as evidenced by the two large and two small phenotypic classes (*Hint 1*). The two largest categories (purple, smooth and green, wooly) are parental types while the two smallest categories (purple, wooly and green, smooth) are recombinant types. To determine the distance between genes, calculate the frequency of recombinant classes. In this example:

Recombination frequency	=	(205 + 195) / (790 + 205 + 195 + 810)
	=	400/2000
	=	0.20 x 100 = 20% or 20 map units

Because a testcross was performed, the phenotype observed corresponds directly with the alleles received from the dihybrid (a dominant phenotype results if dominant alleles were passed to the offspring, a recessive phenotype if recessive alleles are passed to the offspring). The two parental phenotypic classes result from a lack of crossing over between the genes of interest. Therefore, taken together, these classes represent the genotype of the dihybrid female (*Hint 7*). In this case:

Trans Configuration	purple smooth	Aw wo
	green wooly	aw Wo

P6-2. The three genes are linked on a single chromosome as suggested by the lack of a 1:1:1:1:1:1:1:1 ratio resulting from the testcross (*Hint 1*). The two largest categories represent the parentals while the two least frequent classes result from a double crossover (*Hint 2*). Again, distances between genes are determined by calculating recombination frequencies.

- Determine correct gene order by comparing double crossover classes to the parentals (*Hint 3*).

Parentals			Doubles		
k	c	e	k	c	e+
k+	c+	e+	k+	c+	e

  Only the gene in the middle changes in a double crossover. Notice that all mutant traits are together on one chromosome in the parentals. In the double crossovers, ebony is located on the other chromosome. Therefore, body color is the gene in the middle.

- Re-write each phenotypic class in the appropriate order, using symbols.

k	e	c	k	e+	c+
k+	e+	c+	k+	e	c
k	e	c+	k	e+	c
k+	e+	c	k+	e	c+

- Identify each phenotypic class by comparing it to the parentals (*Hint 4*).

k	e	c	*Parentals*	k+	e+	c+
k	e	c+	*SCOS (e and c)*	k+	e+	c
k+	e	c	*SCOS (k and e)*	k	e+	c+
k	e+	c	*DCOS*	k+	e	c+

- Calculate map distances .

<u>Distance between k and e</u>	=	(# singles k-e + doubles / total number offspring) x 100
	=	((98 + 88 + 12 + 14) / 4000) x 100
	=	(212 / 4000) x 100
	=	5.3 map units
<u>Distance between e and c</u>	=	(# singles e-c + doubles / total number offspring) x 100
	=	((128 + 134 + 12 + 14) / 4000) x 100
	=	(288 / 4000) x 100
	=	7.2 map units
<u>Distance between k and c</u>	=	k-e distance + e-c distance
	=	7.2 + 5.3
	=	12.5 map units

P6-3. When looking at this data, you find four equally frequent large categories and four equally small categories. This pattern suggests that only two of the three genes being considered are linked together and the third is independently assorting (*Hint 8*). To determine which genes are linked, compare the phenotypes of the four largest classes.

h	c	M
h	c	m
H	C	m
H	C	M

Notice that hairless and sane are always together as are the alternative characteristics (hairy and crazy). These two traits are therefore linked with mischievous/innocent independently assorting. Elimination of the information about this trait yields a simple two-point linkage map problem.

h c	782 + 815 = 1597
H C	790 + 816 = 1606
h C	197 + 200 = 397
H c	195 + 205 = <u>400</u>
	Total = 4000

Distance: hairless/hairy to crazy/sane ((397 + 400)/4000) x 100 = 19.925 map units

Trihybrid F1 genotype (cis arrangement): $\frac{h \quad c}{H \quad C} \quad \frac{M}{m}$

P6-4. In this problem, you are provided with map distances and asked to work back to the data. To do so, you must remember how each of the map distances provided was calculated.

- Calculate the number of double crossover types expected. Without crossover interference, each crossover event occurs independently. Therefore, we multiply the probabilities of crossovers between each pair of genes.

Probability of crossover between x and y = 0.14
Probability of crossover between y and z = 0.16
Probability of a double crossover = 0.14 x 0.16 = 0.0224
2.24% of total offspring are expected to be double crossover types

- Calculate the expected frequency of single crossover types. Remember that the map distance equals the frequency of single crossovers plus the frequency of double crossovers.

  Probability of single crossover (x and y) types = 14% – 2.24% = 11.76%
  Probability of single crossover (y and z) types = 16% – 2.24% = 13.76%

- The remaining individuals are parental types.

  Parental types = 100 – 2.24 – 11.76 – 13.76 = 72.24%

- Construct data set. Assuming 1000 offspring and a cis allele arrangement in the F1 trihybrid female.

x y	z	362		x y	z+	68
x+ y+	z+	362		x+ y+	z	69
x y+	z+	59		x y+	z	11
x+ y	z	58		x+ y	z+	11

P6-5. In this problem, you are asked to assume that crossover interference is 100%. In this case, the occurrence of the first crossover event completely inhibits the second crossover event. As a result, no double crossover types are obtained (*Hint 9*).

x y	z	350	x+ y	z	70
x+ y+	z+	350	x y	z+	80
x y+	z+	70	x+ y+	z	80

P6-6. Given that the number of double crossovers is reduced by 55%, you must first calculate the number of expected double crossovers. Follow the outline from problem 4 to arrive at an answer.

To calculate the number of expected double crossovers:

From above, expected number of double crossovers given no interference = 2.24%

I = 1 - (# doubles observed / # doubles expected)
0.55 = 1 – (observed frequency / 0.0224)
0..55-1.0 = – (observed frequency / 0.0224)
0.45 = observed frequency / 0.0224
0.45 x 0.0224 = observed frequency
0.01008 = observed frequency   Observed frequency of DCO (55% interference) ≈ **1.0%**

x y	z	355	x y	z+	75
x+ y+	z+	355	x+ y+	z	75
x y+	z+	65	x y+	z	5
x+ y	z	65	x+ y	z+	5

P6-7. To calculate the gene-centromere distance, add all MII asci (2:2:2:2 and 2:4:2) together and divide by the total number of hybrid asci recorded. Divide this number by two.

2 x 0.26 map units = (# 2:2:2:2 asci + # 2:4:2 asci) / 2000
0.52 = (# MII asci) / 2000
0.52 x 2000 = #MII asci
1040 = #MII asci (2:2:2:2 and 2:4:2)
#MI asci = 2000 –1040 = 960 (4:4)

P6-8. For each enzyme, determine which clones produce the enzyme and which do not. Compare the enzyme presence with a chromosome presence/enzyme absence with a chromosome absence. For example, if enzyme a is produced in colony A and B but not C, the chromosome on which the gene resides should be present in colony A and B but not C.

Enzyme Activity	Chromosome Location
human enzyme 1a	4
human enzyme 2a	5
human enzyme 3a	1
human enzyme 4a	3

P6-9. This is an example of the "grandfather" method of estimating recombination from pedigree data. In this problem, the male in generation I expresses trait A while his wife carries the allele for trait B. The female in generation II is heterozygous for both traits. The alleles are in the trans configuration (A b obtained from the female, a B obtained from the male). Of the males in generation III, those expressing only trait A or trait B are parental types. Those expressing both or neither are the result of recombination. Male number 1 and male number 4 are recombinants. Our distance calculation will be 2/6 or ~33 map units.

Answers to General Questions

Q6-1. Crossing over may or may not produce recombinant offspring phenotypes. A crossover occurring outside of the region containing the two genes under study will not be detected in the offspring phenotypes. A double crossover occurring between two genes will also go undetected. Remember that in mapping studies, the frequency of recombinant offspring phenotypes is used to estimate the frequency of crossing over. We can only detect crossovers which separate the genes under study.

Q6-2. Linked genes may appear to assort independently if they are distantly located on the chromosome.

Answers to Even Numbered End of Chapter Exercises, Problems and Critical Thinking Questions

2. The linkage pattern is consistent with X linkage.

$$P_1 \; X^{abe,bis} X^{abe,bis} \text{ (female)} \times X^{+} Y \text{ (male)}$$

$$F_1 \; X^{+} X^{abe,bis} \times X^{abe,bis} Y$$

F_2: Since crossovers can only occur in F_1 females, this is the same as a testcross: both sons and daughters will have the phenotype of the X chromosome inherited from the mother. Thus, in the F_2, both sons and daughters have roughly the same distribution of phenotypes. Abnormal and brown are the recombinant classes (43 + 45 + 37 + 35 = 160). The two loci are 16.0 map units apart on the X chromosome ([160/1,000] × 100).

4. This is a simple three-point cross complicated by the fact that these loci are detected by electrophoretic methods. The alleles are codominant and thus, although we can't do a true testcross (no recessive homozygotes exist), we tested the trihybrid females by crossing them with males that were homozygous for the slow alleles. Thus the cross to produce the data above was as follows:

$$got^f got^s\ amy^f amy^s\ sdh^f sdh^s \times got^s got^s\ amy^s amy^s sdh^s sdh^s$$

In looking at the offspring, we know that each has a slow allele at each locus contributed by the homozygous males. Thus we look at the other allele in each of the offspring to know the alleles contributed by the trihybrid female. By doing that we have the same information we would have in a testcross.

The pattern of numbers among the eight offspring classes is the pattern we are used to for linkage of the three loci. We can tell from the two groups in largest numbers (nonrecombinants—classes 1 and 2) that the alleles are in the coupling (*cis*) arrangement. If we compare either of the nonrecombinant offspring with the double recombinant classes (7 and 8), we can see that the *amy* locus is in the middle. The comparisons of classes 8 with 2 or 7 with 1 should be especially clear. Again, disregard the slow alleles at each locus and look only at the other alleles, the ones contributed by the trihybrid. We can now infer that the trihybrid female had the following chromosomal arrangement:

$$\frac{got^f\ amy^f\ sdh^f}{got^s\ amy^s\ sdh^s}$$

You should now be able to see that classes 3, 4, 7, and 8 have crossovers in the *got–amy* region and classes 5, 6, 7, and 8 have crossovers in the *amy–sdh* region. Tallying these numbers, we see that the recombinants make up (11 + 14 + 1 + 1)/1,000 = 0.027, or 2.7%, in the *got–amy* region and (58 + 53 + 1 + 1)/1,000 = 0.113, or 11.3%, in the *amy–sdh* region. Thus the map distances between loci are 2.7 and 11.3 centimorgans, respectively.

We expect 0.027 × 0.113 = 0.00305 double crossovers, or 3.05 per thousand. We observed only 2. Thus the coefficient of coincidence is 2/3.05 = 0.6557, or about 66% of the expected double crossovers are actually observed.

6. a. The brittle and glossy loci are linked; ragged is assorting independently. The brittle to glossy distance is 8.0 map units (23 + 17 + 21 + 19 recombinants). The offspring of this cross form a pattern of four classes in high numbers, all about equal, and four classes in low numbers, all about equal. This is the pattern of simple linkage of two loci with a third locus assorting independently. There are two ways to approach this problem. By inspection you should be able to see that among the four classes in high numbers, the brittle and glossy mutants are together and their wild-type alleles are also together, with the ragged phenotype acting independently. Thus the ragged locus is assorting independently and the four classes in high numbers are actually two classes with regard to brittle and glossy—they are the parentals. Similarly, the four classes in low numbers are the recombinants. If you cannot see this, then arbitrarily assign one reciprocal set in high numbers as parentals and one reciprocal set in low numbers as double crossovers. You will then calculate map distances such that brittle and glossy are 8 map units apart and ragged is 50 map units from the "middle" locus. In other words ragged is assorting independently.

b. The *bt*1 and *gl*17 alleles are in *cis* (coupling) phase.

c. Not relevant.

8. Notchy should be known on the X chromosome by the results of reciprocal crosses as well as linkage with other X-linked loci. If a trihybrid female is testcrossed (to an ancon, spiny-legged male fly hemizygous for the ny allele), the offspring will consist of, for example:

	Males	Females
Parentals (*an sple* or an^+ $sple^+$)		
ancon, spiny legs, notchy	112	113
ancon, spiny legs	113	112
notchy	112	113
wild-type	113	112
Recombinants (*an* $sple^+$ or an^+ *sple*)		
ancon, notchy	12	13
ancon	13	12
spiny legs, notchy	12	13
spiny legs	13	12

Ancon and spiny legs are linked, separated by 10 map units, whereas notchy segregates independently. Due to the genetic nature of the male parent, the sexes segregate similarly. Thus, of one thousand offspring, nine hundred will be in the parentals group, divided equally among sexes and genotypes. One hundred will be recombinant (10%), again, divided equally among sexes and genotypes.

10. a. Tunicate and Glossy are linked; Ligule assorts independently. b. *TG/tg; L/l.* c. 20.5 map units. We see four majority classes in equal frequencies (58, 55, 53, 59), and four minority classes in equal frequencies (15, 13, 16, 14), indicating that one gene is not linked. If two genes are linked, they should produce a 1:1:1:1 ratio in majority classes. Comparing Tunicate and nontunicate with Liguled and liguleless or Liguled and liguleless with Glossy and nonglossy produces this 1:1:1:1. Therefore, Tunicate and Glossy are linked. The nonrecombinants will appear in the majority classes; Tunicate, Glossy and nontunicate, nonglossy are these classes. The alleles are in the *cis* configuration, *TG/tg.*

$$\text{Map units} = \frac{15+13+16+14}{283} \times 100 = 20.5$$

12. a. 5 wild-type: 1 javelin: 3 ebony: 1 javelin, ebony (no crossing over in *Drosophila* males)

	b.	*c.*
thread, ebony	0.368	0.365
Delta	0.368	0.365
wild-type	0.005	0.004
thread, Delta, ebony	0.005	0.004
thread, Delta	0.110	0.113
ebony	0.110	0.113
thread	0.017	0.020
Delta, ebony	0.017	0.020

a. The distance between javelin and ebony is greater than 50 map units; so, the genes should behave as if they were unlinked in females. The cross is:

$$\frac{jv\ e}{jv^+\ e^+} \times \frac{jv\ e}{jv^+\ e^+}$$

Since there is no crossing over in males, however, the following chart can be drawn:

♀	♂ $jv\ e$	$jv^+ e^+$
$jv\ e$	$jv\ e/jv\ e$	$jv\ e/jv^+ e^+$
$jv^+ e^+$	$jv\ e/jv^+ e^+$	$jv^+ e^+/jv^+ e^+$
$jv\ e^+$	$jv\ e/jv\ e^+$	$jv\ e^+/jv^+ e^+$
$jv^+ e$	$jv\ e/jv^+ e$	$jv^+ e/jv^+ e^+$

b. Calculate distances: $th - Dl = 23$; $Dl - e = 4.5$

Calculate double crossover frequency = (0.23)(0.045) = 0.01035

Calculate $th - Dl$ crossovers = 0.23 = 0.01035 = 0.21965

Calculate $Dl - e$ crossovers = 0.045 - 0.01035 = 0.3465

Nonrecombinants = 1 – (0.01035 + 0.21965 + 0.03465) = 0.73535

Each class within a category will be 1/2 of the above frequencies, since recombination is a reciprocal event. For (c), proceed the same way, except the frequency of double crossovers will be different.

$$\text{coefficient of coincidence} = \frac{\text{observed}}{\text{expected}};\ \text{(coefficient of coincidence) (expected)} = \text{observed}$$

(0.4)(0.01035) = 0.00414. This number is then subtracted from single exchange classes. (It is used above in place of 0.01035.)

14. Single crossover between the loci, involving only two chromatids, should produce a pattern similar to

$$ab,\ ab,\ ab^+,\ ab^+,\ a^+b,\ a^+b,\ a^+b^+,\ a^+b^+$$

If crossing over occurred at the two-strand stage, each crossover would involve both chromosomes and therefore all four chromatids, such as

$$ab^+,\ ab^+,\ ab^+,\ ab^+,\ a^+b,\ a^+b,\ a^+b,\ a^+b$$

A single crossover involves only two chromatids, but complex crossovers can involve three or four chromatids. For example, a three-strand double crossover between the a and b loci will produce a tetratype:

$$ab,\ ab,\ ab^+,\ ab^+,\ a^+b^+,\ a^+b^+,\ a^+b,\ a^+b$$

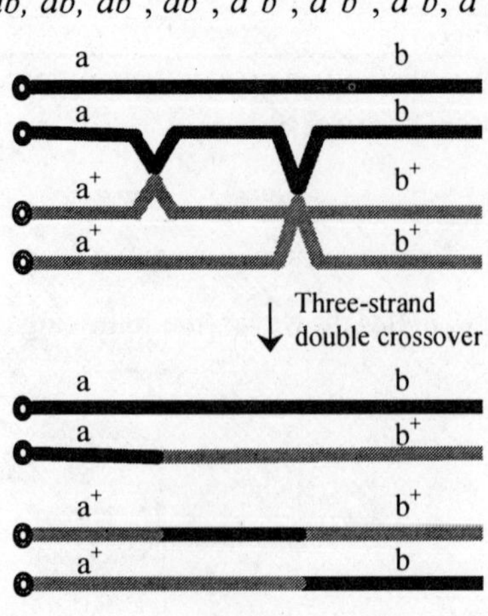

16.

a. PD, 1, 2, 6, 8, 10, 11; NPD, 9; TT, 3, 4, 7; unscorable, 5, 12.

b. Yes, PD > NPD.

c. Map units = ([NPD + {1/2}TT]/Total) × 100 = ([1 + {1/2}3]/10) × 100 = 25. The loci are 25 map units apart.

18. a. FDS, 1, 4–6, 8, 9, 11; SDS, 2, 7, 10; unscorable, 3, 12.

b. The distance between the fuzzy locus and its centromere is (1/2)%SDS = (1/2)30% = 15 map units.

20. $ab \times a^+b^+ \rightarrow ab/a^+b^+$, which undergoes meiosis. Twelve map units means NPD + (1/2)TT = 12% of the total asci. For example, of one hundred asci dissected, six could be NPD and twelve TT.

22. The best first-order estimate is that the *a* and *b* loci are on opposite sides of the same centromere. The *a* locus is 5.5, and the *b* locus is 10.5, map units from the centromere; the *a–b* distance, based on unordered spores is 15 map units, confirming the arrangement.

24. Assume the order of centromere, *a, b*.

Class 1: No crossovers.

Class 2: A four-strand double crossover between the *a* and *b* loci.

Class 3: One crossover between the loci.

Class 4: A two strand double crossover, one between the centromere and the *a* locus and the other between the loci.

Class 5: One crossover between the centromere and the *a* locus.

Class 6: Three crossovers. The first is between the centromere and the *a* locus on chromatids 2 and 3; the second is between the two loci on chromatids 1 and 3; and the third is between the two loci on chromatids 2 and 4.

Class 7: Two crossovers. The first is between the centromere and the *a* locus on chromatids 2 and 3; the second is between the loci on chromatids 2 and 4.

26. a. *a* to centromere: 6 map units; *b* to centromere: 0; *c* to centromere: 13; *d* to centromere: 20.5. b. *b, c,* and *d* are linked.

c.
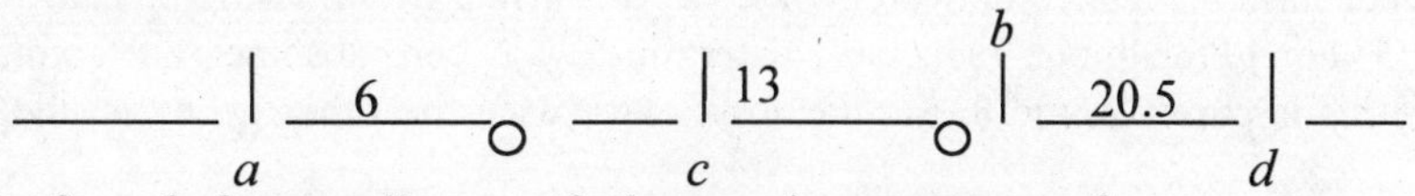

We use the formula for map distance of a locus to its centromere of:

$$\frac{1/2\,(number\ SDS)}{100} \times 100$$

a to centromere = (1/2)(12) = 6 *b* to centromere = 0

c to centromere = (1/2)(26) = 13 *d* to centromere = (1/2)(41) = 20.5

Examine each cross for PD and NPD. The second cross yields PD > > NPD, indicating that *b* and *c* are linked. The third cross indicates that *c* and *d* are linked. We know that *b* is located at its centromere.

Calculate the *c* – *d* distance as (number of NPD + [1/2]number TT)/100 × 100 = (3 + [1/2]50) = 28 map units. If *c* and *d* are on the same side of the centromere, most SDS for *c* should also be SDS for *d*. We see 26 SDS for *c* and 11 for *c* and *d*. Therefore *c* and *d* are probably on opposite sides of the centromere, consistent with the *c* - *d* distance of 28 map units calculated above.

28. You must analyze two genes at a time. Begin by classifying each ascus.

Ascus	*pab*	*pk*	*ad*	*pab-pk*	*pab-ad*	*pk-ad*
I	FDS	FDS	FDS	PDT	PDT	PDT
II	SDS	SDS	SDS	PDT	PDT	PDT
III	FDS	SDS	SDS	TT	TT	PDT
IV	FDS	FDS	SDS	PDT	TT	TT
V	SDS	SDS	SDS	TT	TT	PDT
VI	SDS	SDS	SDS	PDT	TT	TT
VII	SDS	SDS	FDS	PDT	TT	TT
VIII	SDS	SDS	SDS	PDT	TT	TT

We see no NPD, so all three genes are linked. Calculate the gene-to-centromere distances using (1/2 number SDS)/93 ×100.

pab-centromere = (1/2)[(35 + 2 + 2 + 1 + 3)/93] × 100 = 23 map units

pk-centromere = (1/2)[35 + 9 + 2 + 2 + 1 + 3)/93] × 100 = 28 map units

ad-centromere = (1/2)[35 + 9 + 7 + 2 + 2 + 3)/93] × 100 = 31 map units

Now calculate gene-to-gene distances using ([1/2]TT + NPD)/93 × 100.

pab-pk = (1/2)[(9 + 2)/93] × 100 = 6 map units

pab-ad = (1/2)[(9 + 7 + 2 + 2 + 1 + 3)/93] × 100 = 13 map units

pk-ad = (1/2)[(7 + 2 + 1 + 3)/93] × 100 = 7 map units

All three genes are on the same side of the centromere.

30. The +/– patterns seen under the columns headed chromosome 6, 14, X, respectively.

32. Only the type A, fructose-intolerant children are recombinant. Thus, estimate θ at 0.2 (2 of 10 = 20%). Probability of a particular nonrecombinant child = 0.8/2 = 0.4. Probability of a particular recombinant child = 0.2/2 = 0.1. Z = $\log([0.4]^8[0.1]^2/[0.25]^{10})$ = log(6.872) = 0.837. This value supports linkage.

34. glutathione reductase: 8; malate dehydrogenase: 2; adenosine deaminase: 20; galactokinase: 17; hexosaminidase: 5. For gluthathione reductase, determine what chromosomes are common to both clones A and B—this is chromosome 8. Similar logic will place the other genes on the appropriate chromosomes.

36. Lines A and C. We need to determine the lines that have chromosome 3; these lines are expected to be positive for the enzymes; we see that lines A and C have chromosome 3.

Critical Thinking Question

2. If 4% of all tetrads have a single crossover during meiosis, then 2% of all chromatids will be recombinant (half the chromatids in 4% of the cells). Thus, in fruit flies, the distance between the loci will be 2 map units. In *Neurospora,* 4% of asci will exhibit second division segregation, which is also 2 map units between loci. In yeast, a single crossover will generate a tetratype only, not a nonparental ditypc. Again, the distance between the two loci will be recorded as 2 map units.

CHAPTER 7
LINKAGE AND MAPPING IN PROKARYOTES AND BACTERIAL VIRUSES

Key Genetic Concepts

Bacterial Culturing and Screening Techniques: Bacteria are the model organisms of choice for genetic analysis of biochemical traits. Strains with nutritional requirements can be generated and studied to determine which genes in a biosynthetic pathway are nonfunctional or defective. A bacterial strain which is able to grow on a medium containing only the minimal nutritional necessities (**minimal medium**) is referred to as a **prototroph**. Strains of bacteria with a defect in a biosynthetic pathway have additional nutritional requirements and are referred to as **auxotrophic**. Auxotrophic strains can grow only on media containing the nutrient they are unable to synthesize. Mutant strains that can grow on a supplemented or **enriched medium** are called **conditional-lethal mutants**. Various kinds of **selective media** (minimal media plus one additional substance) are used to identify conditional-lethal mutants. Bacteria may also differ in their **resistance** and **sensitivity** to antibiotics. Drug sensitivity can be use to select colonies with particular mutations. **Replica-plating** is a technique which is used to determine if a bacterial strain is auxotrophic.

Bacterial Sexual Processes: Bacteria possess sexual processes which allow genetic studies to be conducted. Three different types of processes that result in the combining of genetic material from two individuals occur.

- The first, **transformation**, involves the uptake of extraneous DNA found in the environment and the incorporation of this piece of genetic material into the cell's genome through recombination. Cells that undergo transformation contain a surface protein called a **competence factor** which is needed to bind the extra-cellular piece of DNA. Two crossovers are needed to incorporate a piece of the foreign DNA into the cell's genome. The remaining exogenous DNA is degraded.
- **Conjugation** in bacteria such as *E. coli* involves the transfer of genetic material from one cell (the donor) to another (the recipient). Cells carrying a **fertility** or **F factor** (F^+ strain) are capable of transferring a single strand of DNA to a recipient cell lacking an F factor (F^- strain). The F factor is an independently replicating piece of DNA called a **plasmid**. In another strain, **Hfr**, the F factor is part of the bacterial chromosome. During conjugation, genetic material from the donor cell passes to the recipient cell through **F-pili** or **sex-pili** which form a connection between the two cells. The foreign DNA is incorporated into the genome of the recipient through the process of crossing over.
- The final sexual process in bacteria is **transduction**. In this case, a bacterial virus or **bacteriophage** transfers bacterial DNA between cells. The bacteriophage replicates within a bacterial host cell. An occasional error results in the incorporation of bacterial DNA into the virus. When this virus infects another cell, bacterial DNA carried by the virus may be recombined into the new host.

Sexual Processes and Mapping the Bacterial Chromosome: Each of the sexual processes discussed above involve the transfer and incorporation of DNA from one bacterial cell into a second cell. These processes can therefore be used to map the location of bacterial genes.

- Transformation mapping provides a measure of the relative distance between two loci called a **co-occurrence** or **cotransfer index**. A large co-occurrence index indicates that the genes under study are close together while a small co-occurrence index indicates a greater map distance. The number of recipient bacteria that incorporate one or the other (single transformants), or both (double transformants) of the donor alleles is counted. The cotransfer index is calculated by dividing the number of double transformants observed by the total number of transformants observed.
- The processes of conjugation can be exploited to map the relative positions of genes on the bacterial chromosome. During conjugation, bacterial genes are transferred sequentially over time with the F factor being transferred last. An **interrupted mating** technique can be used to map genes. At specific time intervals, "mating" cells are disrupted, ending the transfer of genetic material. The genotypes of the F^- cells are then determined by replica plating. This technique provides estimates, in units of time, of the distance between genes and is most accurate for genes that are distantly located.
- Transduction can also be used to establish gene order and map distances between bacterial genes. Three-factor transduction experiments are the most useful in mapping studies. A prototroph is used to make transducing phages which are then provided with auxotrophic host cells. The number of cells that are transduced for one, two, or three loci is determined. **Cotransductance** values are then to establish relative distances between genes.

Problem-Solving Hints

Hint 1: The bacterial chromosome is circular rather than linear as in eukaryotic cells.

Hint 2: The F factor integrates into the chromosome at different places in different bacterial strains of the same species. The direction of transfer is consistent within a strain but may differ between strains (see Table 7.4 in Tamarin). For example, strain A may transfer genes in the following order: A B C D E while strain B transfers gene D first, followed by C B A and finally E.

Hint 3: If two genes are close together, a high frequency of cotransfer or cotransductance will be observed. A low frequency of cotransfer or cotransductance indicates that the genes are distant. Remember that the cotransfer/cotransductance index is a measure of how often two genes are transferred together: it is the opposite of a recombination frequency.

Hint 4: Gene order can be established in transformation and transduction experiments by examining the genotype of the least frequent class. The gene in the middle is the only gene that is *not* transferred to the recipient cell. For example, donor genotype a^+ b^+ c^+ is used to transform recipient genotype a^- b^- c^-. The least frequent class has the genotype a^- b^+ c^+. The gene order is b a c or c a b.

Hint 5: Double transformant transduction classes are those in which two of the donor markers have been transferred to the recipient cell. Single transformant/transduction classes are those in which only one of the donor markers has been transferred to the recipient cell. Cells that were not transformed/transduced are not considered in mapping experiments.

Hint 6: Conjugation is used to determine relative map distances. Remember that distances obtained from interrupted mating studies are expressed in time. This technique is most effective for genes that are distantly located on the chromosome.

Practice Problems

Problem 7-1. The gene order of several Hfr strains was determined by means of interrupted mating. Using this information, construct a map of the bacterial chromosome. Indicate the location of each F factor.

Strain	Order of Transfer							
1	C	O	N	J	U	G	A	T
2	U	G	A	T	I	0*	N*	C
3	T	I	O*	N*	C	O	N	J
4	N	J	U	G	A	T	I	O*
5	G	U	J	N	O	C	O*	N*

Problem 7-2. A prototrophic strain (cys$^+$ leu$^+$) was used to transform an auxotrophic strain (cys leu$^-$). Examine the patterns of bacterial colony growth on the different media plates diagrammed below. Identify the genotype of each colony. Which are transformed cells and which are untransformed cells?

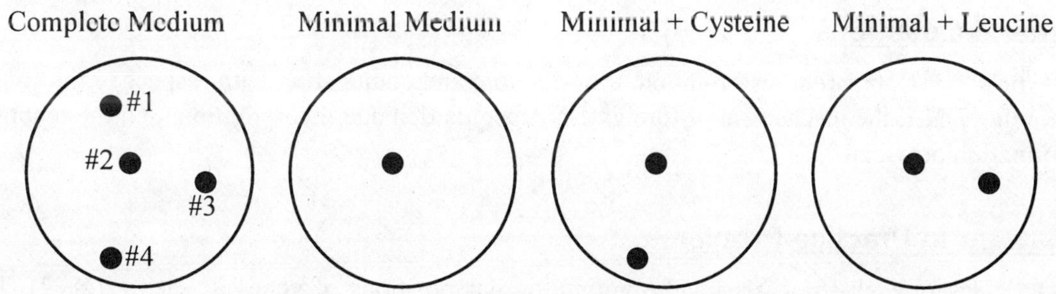

Problem 7-3. A bacterial cell with genotype a$^+$b$^+$ served as the donor strain in a transformation experiment. The recipient strain had the genotype a$^-$b$^-$. The transformed classes obtained included: 400 a$^+$b$^+$; 250 a$^+$b$^-$ and 304 a$^-$b$^+$. What is the cotransfer index?

Problem 7-4. A transduction experiment was conducted using the donor strain a$^+$b$^+$c$^+$ and the recipient strain a$^-$b$^-$c$^-$. The a$^+$ allele was used as the selection variable. From the transductant classes obtained, determine the cotransductance frequencies.

a$^+$b$^+$c$^+$	30	a$^+$b$^+$c$^-$ 44
a$^+$b$^-$c$^-$	60	a$^+$b$^-$c$^+$ 10

Problem 7-5. An interrupted mating experiment was conducted with the donor Hfr strain $a^+b^+c^+d^+e^+f^+g^+$ and the recipient strain $a^-b^-c^-d^-e^-f^-g^-$. The appearance of donor genetic markers in recipient cells is given below. What is the gene order and distance between genes?

Time	a^+	b^+	c^+	d^+	e^+	f^+	g^+
5 mins	--	--	+	+	--	--	--
10	--	--	+	+	--	--	--
15	--	--	+	+	+	--	--
20	--	--	+	+	+	--	--
25	--	--	+	+	+	--	--
30	+	--	+	+	+	--	--
35	+	--	+	+	+	+	--
40	+	--	+	+	+	+	+
45	+	--	+	+	+	+	+
50	+	+	+	+	+	+	+

Problem 7-6. A $lys^+his^+val^+$ strain is used to transduce a $lys^-his^-val^-$ strain. Transductants were selected for synthesis of lysine by plating cells on minimal media. The genotypes found among the transformants are given below. Determine the gene order and relative cotransductance frequencies. Which of the cotransductance frequencies represents the greatest distance between genes?

$lys^+his^-val^-$	800	$lys^+his^-val^+$	10
$lys^+his^+val^-$	450	$lys^+his^+val^+$	200

General Questions

Question 7-1. Compare transformation, transduction, and conjugation with respect to the source of foreign DNA, the mechanism of foreign DNA acquisition and incorporation, and the mapping information provided.

Answers to Practice Problems

P7-1. Diagram each strain separately, noting similar positions of genes in each (*Hint 2*). The symbol [F] is used to denote the position of the F-factor.

CO	N	J	U	G	A	T	[f]		
C[F]			U	G	A	T	I	O*	N*
CO	N	J	[F]			T	I	O*	N*
	N	J	U	G	A	T	I	O*	
CO	N	J	U	G	[F]			O*	N*

Order: **CONJUGATION**

P7-2. All colonies grow on complete medium but only colony #2 is able to grow on the minimal medium. Colony #2 is prototrophic while colonies #1, 3, and 4 are auxotrophic. Examination of the selective medium plates indicates that colony #4 grows in the presence of cysteine indicating this colony lacks the gene for cysteine synthesis. Colony #3 grows in the presence of leucine indicating that the gene for leucine synthesis is defective in this colony. Colony #1 is unable to grow on either of the selective media plates indicating that it has both defective genes.

Colony	Genotype	Transformed?
1	$cys^- \ leu^-$	NO
2	$cys^+ \ leu^+$	YES
3	$cys^+ \ leu^-$	YES
4	$cys^- \ leu^+$	YES

P7-3. The cotransfer index is calculated by dividing the number of double transformants by the total number of transformants (*Hint 5*).

$r = \#a^+b^+$ transformants / total # transformants
$r = 400 / (400+250+304)$
$r = 400 / 954 = 0.4193$ or ~42%

P7-4. In this problem, marker a^+ is used as the selection variable. We can therefore calculate only the $a^+ b^+$ and a^+c^+ cotransductance indexes.

$a^+ b^+$ cotransductance frequency = # $a^+ b^+$ transductants / total transductants obtained
= (30 + 44) / 144
= 0.514 or ~51%

$a^+ c^+$ cotransductance frequency = # $a^+ c^+$ transductants / total transductants obtained
= (30 + 10) / 144
= 0.278 or ~28%

P7-5. To determine the gene order, examine the time chart provided (*Hint 6*). The first gene transferred (earliest time) is the first gene in the sequence while the last to be transferred is the last gene. The distance between genes can be calculated by subtracting the time of the previous entry from the time of the next gene entry. Notice that genes c and d cannot be ordered because they were both transferred within the first five minutes.

Gene	Time first observed
a^+	30
b^+	50
c^+	5
d^+	5
e^+	15
f^+	35
g^+	40

Gene order: c^+,d^+ ———— e^+ —— a^+ —— f^+ —— g^+ —— b^+
Distances: <5 10 15 5 5 10

P7-6. Notice that only the cells prototrophic for lysine are reported. To determine the gene order, examine the class with the lowest frequency: $lys^+his^-val^+$. Notice that his^+ was not transferred indicating that it is the middle gene (*Hint 4*). The gene order is lys his val. Calculate cotransductance indexes as before. Note that all distances must be in reference to lysine.

$lys^+ val^+$ cotransductance frequency = # $lys^+ val^+$ transductants / total transductants obtained
= (10 + 200) / 1460
= 210 / 1460 = 0.1438 or 14.38%

$lys^+ his^+$ cotransductance frequency = # $lys^+ his^+$ transductants / total transductants obtained
= (450 + 200) / 1460
= 650 / 1460 = 0.4451 or 44.51%

The 0.1438 cotransductance frequency represents the greatest actual distance between genes. The further apart two genes are the less likely the are to be cotransduced (*Hint 3*).

Answers to General Questions

Q7-1.

Characteristic	Transformation	Transduction	Conjugation
Source of DNA	exogenous in environment	phage	another bacterial cell
Mechanism of foreign DNA acquisition	competence factor on recipient cell surface	infection by phage	donor with F factor, cell to cell contact mediated by sex-pili
Mechanism of foreign DNA incorporation	crossing-over	crossing-over	crossing-over
Mapping information	gene order and map distance	gene order and map distance	gene order and *relative* distance

Answers to Even Numbered End of Chapter Exercises, Problems and Critical Thinking Questions

2. A heterotroph requires an organic source of carbon; an auxotroph has a particular nutritional requirement. A minimal medium contains only those substances required by the wild-type organism to grow; a complete medium is minimal medium that has been enriched with a complete array of organic compounds including amino acids and nucleic acid subunits. An enriched medium is the minimal medium to which nutrients have been added. A selective medium is a particular case of an enriched medium in which only one or a few items have been added.

4. a. pen^r; b. azi^s; c. his^-; d. gal^-; e. glu^+; f. ton^+.

6. The original strain was $arg^- his^-$. Colonies on minimal medium + histidine are $arg^+ his^-$. Colonies on minimal medium + arginine are $arg^- his^+$. Colonies on both are wild-type ($arg^+ his^+$).

8. Colony 1, $glu^- xyl^+ arg^-$; colony 2, $arg^- his^-$; colony 3, prototrophic; at least glu^+ (may be unable to use other sugars); colony 4, arg^-; colony 5, not just $thr^- val^-$ or thr^- or val^-. May be ilv^- or ile^- or $thr^- ile^- ilv^- val^-$ or some combination; colony 6, *glu* gal^+ *his* or *glu* $gal^+ his^+$.

10. A plasmid is an independent genetic entity within a cell, usually a small auxiliary circle of DNA. It integrates by a single crossover between the plasmid and the bacterial chromosome. It leaves by a reverse process of "looping out" (see figs. 7.18 and 7.19).

12. When the drug sensitivity locus passes into the F^- strain, there is a general decline in the recovery of recombinants after selection because both the Hfr and F^- members of a conjugation event can be killed.

14. The order is *trp purB pyrC* since the class in lowest numbers (4) represents the double recombinants and hence exchanged the outside loci but not the middle one (*purB*). The relative co-occurrence of *trp* and *purB* is $r = (86 + 67)/171 = 0.89$. The relative co-occurrence of *trp* and *pyrC* is $r = (86 + 4)/171 = 0.53$.

16. *lac* to *gal*, 9 minutes; *gal* to *his*, 27 minutes; *his* to *argG*, 24 minutes; *argG* to *xyl*, 11 minutes; *xyl* to *ilv*, 4 minutes; *ilv* to *thr*, 17 minutes; *thr* to *lac*, 8 minutes. For every interruption, a complete medium plate with selecting antibiotics to kill Hfr cells and one each of the seven selected plates are needed.

18. The class in lowest frequency ($trp^+\ his^-\ tyr^+$, 107) identifies *his* as the middle locus (*trp his tyr*); relative co-occurrences: (Let Total = 2,600 + 418 + 685 + 1,180 + 107 + 3,660 + 11,940):

$$trp \text{ to } his : r = \frac{1{,}180 + 11{,}940}{\text{Total}} = \frac{13{,}120}{20{,}590} = 0.64$$

$$his \text{ to } tyr : r = \frac{3{,}660 + 11{,}940}{\text{Total}} = \frac{15{,}600}{20{,}590} = 0.76$$

$$trp \text{ to } tyr : r = \frac{107 + 11{,}940}{\text{Total}} = \frac{12{,}047}{20{,}590} = 0.59$$

20. *acb.* Try a gene order and draw the exchange(s) necessary to produce $a^+b^+c^+$. Remember that the least frequent class is the double crossover. If the gene order is *abc,* only a single set of exchanges is needed to produce the class $a^+b^+c^+$. Try *abc:*

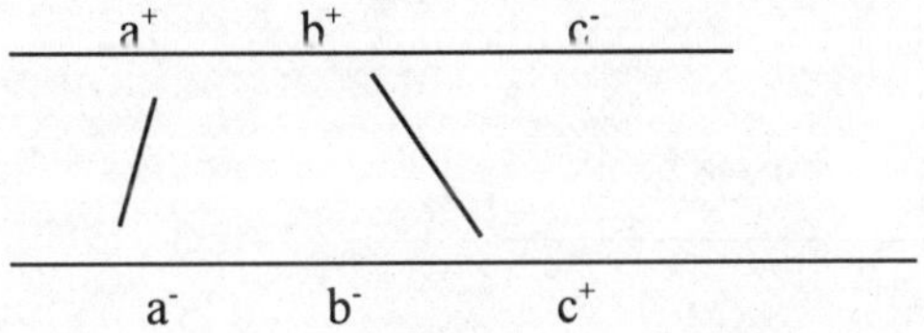

This yields $a^+b^+c^+$ if *b* is in the middle.

Therefore, *b* is not in the middle. Try *c* in the middle:

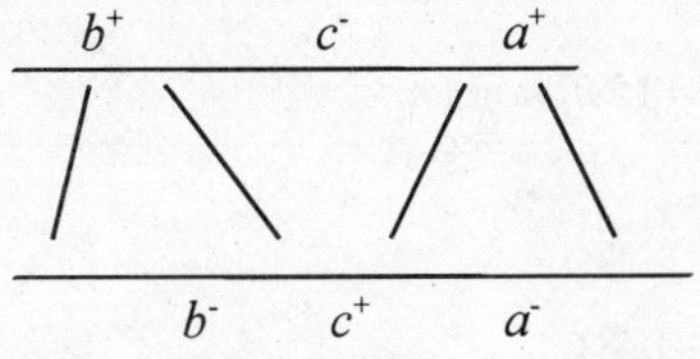

This order produces the desired class as a consequence of four crossovers, which switch the middle locus.

22. a. *arg trp leu* or *trp arg leu.* Since 40% are $trp^+\ arg^+$, *arg* and *trp* must be closer than *trp* and *leu.*

b. Select for *leu,* then score $leu^+\ arg^+$ and $leu^+\ trp^+$. If the first order is correct, $leu^+\ arg^+ < leu^+\ trp^+$. Since *trp* and *arg* are cotransduced 40% of the time, they must be close and adjacent. If the order is *arg trp leu,* the genotype $leu^+\ arg^+\ trp^-$ requires two sets of exchanges, whereas $leu^+\ trp^+$ requires only one set of crossovers. If the order is *trp arg leu,* $leu^+\ arg^+$ requires only one set of exchanges and should be more frequent than $leu^+\ trp^+$.

24.

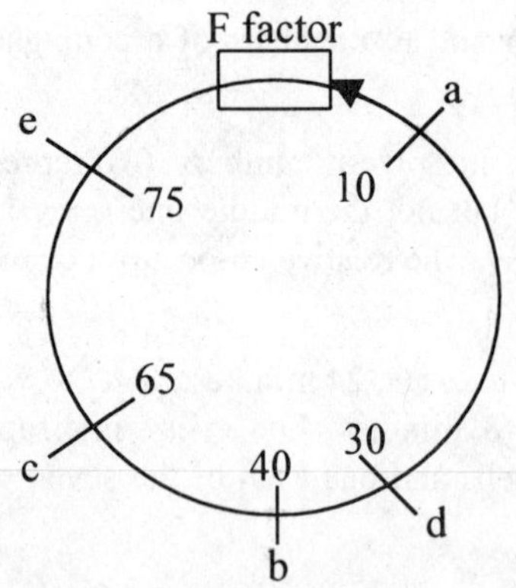

Based on the time of first appearance of the markers, the order is origin *a d b c e.* Let the F factor be located at position 0. Gene *a* appears after only 10 minutes. Therefore *a* is 10 minutes from the origin. Similar logic allows the placement of the other genes.

26. Prophage: an integrated phage. Lysate: material released from a lysed cell. Lysogeny: a bacteria-phage interaction in which the phage has become a prophage that can be induced at a later date. Temperate phage: one capable of becoming a prophage.

28. *m r tu: m-r,* 12.8% = ([162 + 172 + 474 + 520]/10,342)(100); *r-tu,* 20.8% = ([162 + 172 + 853 + 965]/10,342)(100); coefficient of coincidence = (observed double crossovers)/(expected double crossovers) = 334/275.3 = 1.21. The problem is done just like a eukaryotic mapping problem (e.g., *Drosophila*). The data are equivalent to offspring from a testcross (really just the gametes from the trihybrid). For example, the 3,467 and 3,729 classes are nonrecombinant; the 162 and 172 classes are double crossovers.

30.

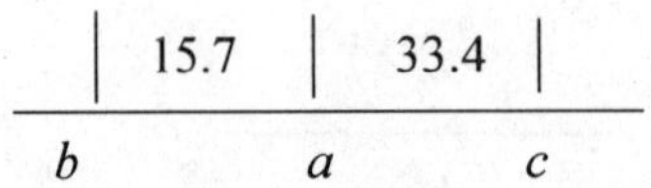

Treat this cross like a three-point eukaryotic cross and group reciprocal classes together. Determine a^+bc and ab^+c^+ as the double crossovers. Since double crossovers rearrange the middle alleles, *a* is in the middle. Proceed to calculate map distances between *b* and *a* as

$$\frac{(371+309+160+178)}{6502} \times 100 = 15.7 \textit{ map units}$$

Similarly,

$$\textit{map distance of } a-c = \frac{(954+879+160+178)}{6502} \times 100 = 33.4 \textit{ map units}$$

32. Specialized transduction comes about through incorrect looping out of a prophage (see fig. 7.25). Generalized transduction involves the inclusion of a piece of bacterial DNA in a phage coat.

34. *thr ara-2 ara-1 ara-3 leu*. Examine those crosses that yield a large difference between ara^+ leu^+ and ara^+ thr^+: crosses 1, 4, and 6. Begin with cross 1 and arbitrarily choose a gene order for the two mutants; for example, assume the order is *thr ara-1 ara-2 leu*. Cross 1 can then be diagrammed as

$$\frac{thr^+ \ ara-1^+ \ ara2^- \ leu^+}{thr^- \ ara-1^- \ ara-2^+ \ leu^-} \qquad \begin{matrix}\text{Donor}\\ \text{Recipient}\end{matrix}$$

Draw the crossovers needed to produce ara^+ leu^+ and ara^+ thr^+:

We see that with the preceding order, ara^+ leu^+ results from two sets of exchanges (b's) and are thus equivalent to a double crossover. Since double crossovers are rare, such transductants should also be rare. The ara^+ thr^+ class would come about by one pair of crossovers (a's). This result is not seen; ara^+ $leu^+ \ll ara^+$ thr^+. Therefore, our initial order is wrong, and the correct order should be *thr ara-2 ara-1 leu*. If you diagram the second order, the numbers are consistent. In cross 4, by similar logic, ara^+ leu^+ $\ll ara^+$ thr^+, so order is *thr ara-1 ara-3 leu*. Since we know *ara-1* is closer to *leu* than *ara-2*, and that *ara-3* is closer to *leu* than *ara-1*, the order becomes *thr ara-2 ara-1 ara-3 leu*. Cross 6 confirms that *ara-3* is closer to *leu* than *ara-2*.

Critical Thinking Question

2. It may seem odd that bacteria have evolved the ability to take up DNA from the environment when that DNA came from a dead bacterium. However, we believe that many processes have evolved because of their ability to generate variability in an organism, variability that might make an organism more capable of adapting to a changing environment. This is especially true for haploid organisms that generally accrue variability by mutation. Thus taking up foreign DNA could be a quick way to generate a lot of variability. Also, this DNA may not be "bad" DNA, even though arising from dead organisms. In fact, DNA may only be available in the environment if there is a lot of it. That would indicate an environment in which there is a lot of bacterial growth. That would indicate that the DNA is from successful organisms rather than failed ones.

CHAPTER 8
CYTOGENETICS

Key Genetic Concepts

Alterations of Single Chromosomes: Various agents such as ionizing radiation, physical stress, and chemical compounds are capable of damaging chromosomes. Breaks can occur in the chromatid which can result in a fragment containing the centromere (**centric**) and a fragment lacking a centromere (**acentric**) if rejoining does not occur. During mitosis, the acentric fragment is often lost. Occasionally, two centric fragments join forming a **dicentric** chromosome containing two centromeres. During nuclear division, the dicentric fragment breaks as it is pulled apart by the spindle fibers. Multiple breaks in a chromosome can also occur. In some cases, a piece of the chromosome is lost (a **deletion**), **duplicated** or reattached in the wrong orientation (an **inversion**).

Alterations Involving Nonhomologous Chromosomes: Breaks occurring simultaneously in two non-homologous chromosomes can be rejoined incorrectly. For example, the ends of two different chromosomes can be joined together resulting in a **reciprocal translocation**. The individual possessing a reciprocal translocation has a normal complement of genetic material and is able to carry out mitosis normally. However, the organism is unable to produce normal gametes.

Variation in Human Chromosome Number: Variations in chromosome number can be detrimental to the organism. In human beings, at least 50% of miscarriages are believed to result from chromosomal abnormalities. **Euploid** cells/individuals may have additional or missing chromosomes sets while **aneuploid** cells/individuals have additional or missing chromosomes. Aneuploidy often results from **nondisjunction** during meiosis. Cells may contain three complete sets of chromosomes (**triploid**), four complete sets of chromosomes (**tetraploid**) or more. Individuals may also be made up of several cell lines, each with a different number of chromosomes (**mosiacs**). Aneuploidy leads to several medically recognized syndromes in human beings including Down Syndrome (Trisomy 21), Edward Syndrome (Trisomy 18), Patau Syndrome (Trisomy 13), Turner Syndrome, the XYY and Triple-X Karyotype, and Klinefelter Syndrome. Diploid cells can be missing a single chromosome (**monosomic**), both copies of a chromosome (**nullisomic**), or they have an extra chromosome (**trisomic**).

Rearrangements of Human Chromosomes: Several human medical syndromes and abnormalities result from chromosome deletions and translocations. Examples include Fragile-X Syndrome (expanding duplication), Cri du Chat Syndrome (deletion), and Familial Down Syndrome (translocation).

Polyploidy: Polyploidy may result from additional sets of chromosomes from the same species (**autopolyploidy**) or from the hybridization of material from two different species (**allopolyploidy**). Although polyploidy results in difficulty during meiotic segregation when an odd number of chromosome sets such as 3N or 5N is involved, it is observed in many plants and some animals.

Problem-Solving Hints

Hint 1: Become familiar with the nomenclature used to describe human chromosome complements. The standard "short-hand" representation of a karyotype includes information about the individuals sex, total number of chromosomes, and the abnormality observed. Specifically, the total number of chromosomes is listed first, followed by a description of the sex chromosomes, and finally a description of the autosomal abnormality, if any. For example, a female zygote with an extra copy of chromosome number 14 would be described as 47, XX, +14.

Hint 2: Information about the phenotype of parent and offspring can be used to determine which parent contributed a gamete with the incorrect number of chromosomes. This is especially true for X-linked traits. For example, knowledge of the parent phenotypes would be helpful when trying to identify which parent contributed the single X possessed by a colorblind, Turner syndrome female.

Hint 3: Nondisjunction can occur during mitosis, during meiosis I, during meiosis II, or during both meiosis I and meiosis II. The likelihood of producing an abnormal gamete will depend upon when nondisjunction occurred. If nondisjunction occurs during meiosis I, all of the gametes will be unbalanced. If nondisjunction occurs during meiosis II, approximately one-half of the gametes will be normal and one-half will be unbalanced.

Practice Problems

Problem 8-1. Provide the karyotype description for the following syndromes: Down Syndrome, Edward Syndrome, Patau Syndrome, Klinefelter Syndrome, Turner Syndrome, Cri du Chat Syndrome, and the "Criminal Karyotype."

Problem 8-2. Diagram all types of nondisjunction of the sex chromosomes that can occur during meiosis I or meiosis II in a normal human female and in a normal human male. Indicate the sex chromosome complement of a zygote resulting from the fusion of each gamete type produced and a normal gamete from the other parent.

Problem 8-3. A man with brown teeth (an X-linked dominant trait) marries a woman with normal white teeth. They have a son with Klinefelter syndrome. Assume that the parents are karyotypically normal and that the boy's abnormal karyotype resulted from non-disjunction. How could you use information about the color of the boy's teeth to determine which parent experienced nondisjunction?

Problem 8-4. Consider the chromosomes below (ovals represent centromeres). Indicate the gene order, centromere position, and banding pattern found on the resulting chromosomes if the area between the arrows is (a) deleted, (b) duplicated, or (c) inverted.

General Questions

Question 8-1. In Chapter 5, you learned that calico/tortoiseshell cats are usually female and that the color pattern results from X-inactivation. It has been noted that calico *males* also occur, but at a very low frequency (1/3,000 cats). To what do these males owe their calico/tortoiseshell coloration? If you were to karyotype one of these individuals, what would you find?

Solutions to Practice Problems

P8-1. (*Hint 1*)

Down Syndrome	47, XX, +21 or 47, XY, +21
Edward Syndrome	47, XX, +18 (majority) or 47, XY, +18
Patau Syndrome	47, XX, +13 or 47, XY, +13
Klinefelter Syndrome	47, XXY
Turner Syndrome	45,X
Cri du Chat Syndrome	46, XX, 5p- or 46, XY, 5p-
Familial Down Syndrome	46, XX, t(14p-, 21q+) or 46, XX, t(14p-, 21q+) (ex)
XYY Karyotype	47, XYY

P8-2. (a) **In females:** (*Hint 3*) **In males:**

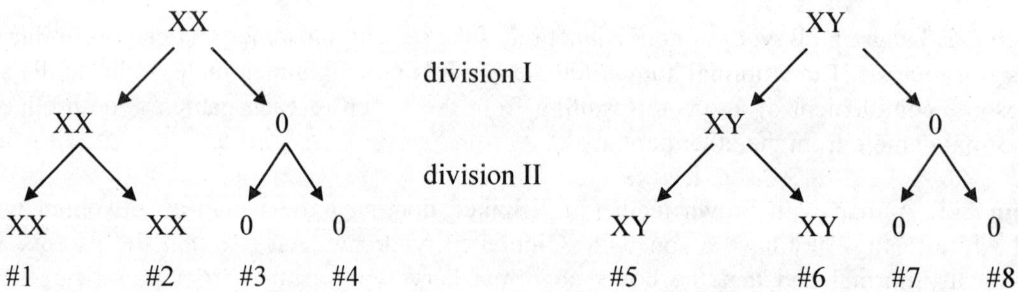

Gamete Number	Type	Normal Gamete	Zygote
#1 and #2	Disomic for X	X or Y	XXX or XXY
#3 and #4	Nullisomic for X	X or Y	X0 or 0Y
#5 and #6	Disomic for sex	X	XXY
#7 and #8	Nullisomic for sex	X	X0

(b) In females: **In males:**

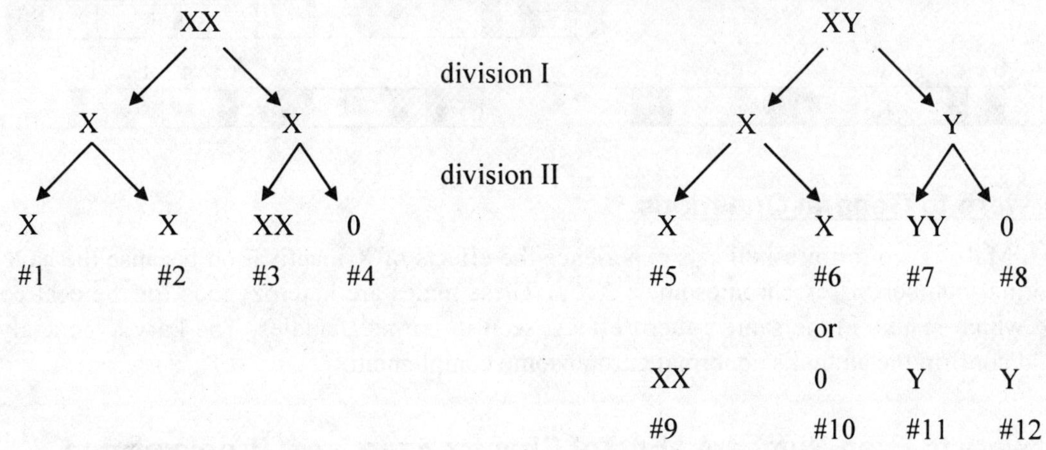

Gamete Number	Type	Normal Gamete	Zygote
#1 and #2	Normal X	X or Y	XX or XY
#3	Disomic for X	X or Y	XXX or XXY
#4	Nullisomic for X	X or Y	X0 or 0Y
#5 and #6	Normal X	X	XX
#7	Disomic for Y	X	XYY
#8 and #10	Nullisomic for sex	X	X0
#9	Disomic for X	X	XXX
#11 and #12	Normal for Y	X	XY

P8-3. The sex chromosome complement of a Klinefelter male is XXY. Nondisjunction in the mother or father could produce this complement. We know that the Y chromosome was obtained from the father. However, it is unclear if the father also donated an X chromosome to his son (primary nondisjunction) or if the mother donated two copies of her X chromosome to her son (primary or secondary nondisjunction). The color of the males teeth can provide clues as to which parent nondisjunction occurred in. Remember that brown teeth is an X-linked dominant trait. In our example, the female is normal and does not possess the allele for brown teeth but her husband does. If the male contributed both his X^B and Y chromosomes, the Klinefelter child should have brown teeth (X^BX^bY). If, on the other hand, the female contributed two X chromosomes, the Klinefelter child should have white teeth (X^bX^bY) (*Hint 2*).

P8-4.

(a)

(b)

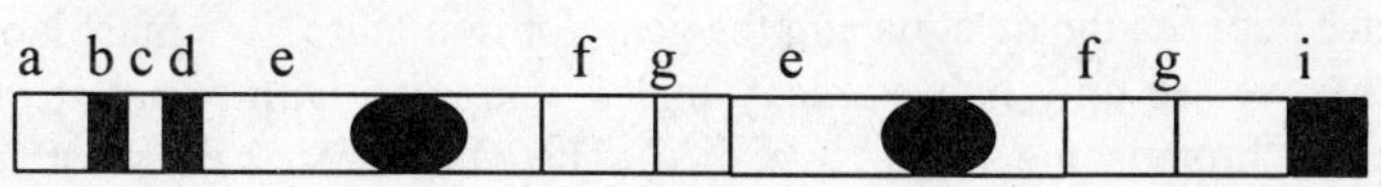

(c)

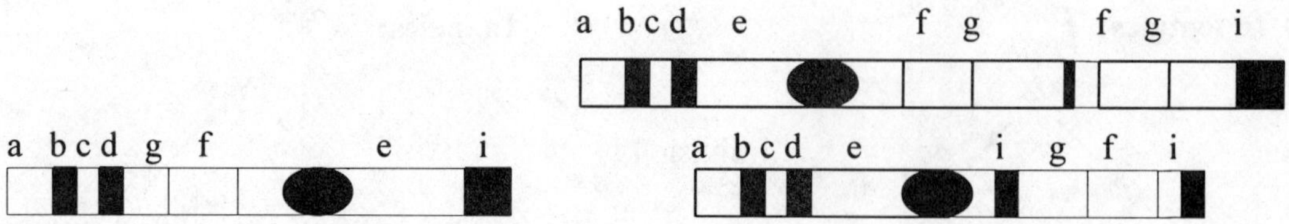

Answers to General Questions

Q8-1. Male calico/tortoiseshell cats experience the effects of X-inactivation because the have an abnormal number of sex chromosomes: XXY. These males are heterozygous for the coat color gene which results in the same pattern effects seen in normal females. The karyotype analysis would confirm the animal's abnormal chromosome complement.

Answers to Even Numbered End of Chapter Exercises, Problems and Critical Thinking Questions

2. Yes, if the deletion moves a gene next to heterochromatin.

4. Any arrangement will suffice that will result in an abnormal chromosomal complement in a gamete, including heterozygous inversions and translocations.

6. One normal chromosome; one normal inversion chromosome; one dicentric chromosome with duplications and deficiencies; one acentric chromosome with duplications and deficiencies (figure following).

8. Inversion (some effects occur only in the heterozygous condition). Look for a loop at meiosis or in salivary gland chromosomes.

10. We have five heterozygous genes, so we expect to see thirty-two genotypes, but we see only six genotypes. We see no exchanges between genes B, C, and D. These three genes could be so tightly linked that no recombination occurs between them (unlikely). Genes B, C, and D could be within an inversion that is heterozygous in the heterozygous parent. The recombination that does occur within this inversion results in inviable zygotes.

12. We see about half the enzyme activity in crosses with strains B and C. Therefore the gene must be located in the region that is common to both strains, approximately in the region located 25 to 35 map units from the left end.

14.

$$\frac{v \qquad\qquad m \qquad\qquad s}{/////////////////////\ 1.1 \qquad m^{+} \qquad\qquad s^{+}}$$

The use of X-rays alerts us to chromosomal aberrations. The fact that a vermilion female appears when we expect all wild-type females indicates that we have a deletion. The deletion must end between *v* and *m*. If the deletion included miniature, we should have seen a vermilion, miniature female. We can draw the chromosomes of the flies in the second cross as

$$\frac{v \qquad m \qquad s}{///////////// \qquad m^{+} \quad s^{+}} \times \underline{v \qquad m \qquad s}$$

The question is, What is the distance between the deletion and the gene for miniature? We must look at the males from the cross. Note that we see only half as many males compared with females; those males that received the deletion X chromosome must have died. The vermilion male must have resulted from a recombination between the end of the deletion and miniature:

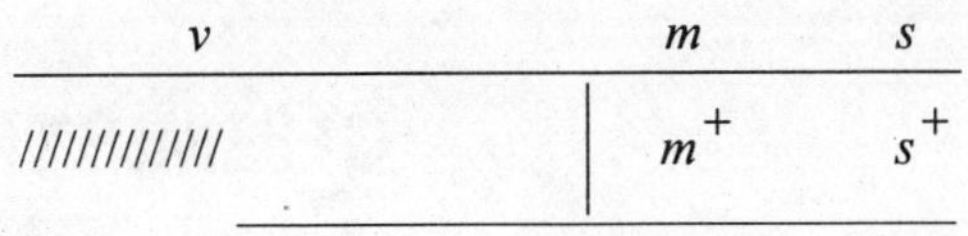

yields the following chromosomes:

v m + s + and /////////////// m s

Therefore, map distance = (vermilion males/total males) × 100 = 1/90 × 100 = 1.1 map units.

16. 3→ 4→ 1→ 2. Inversion of u p o will yield (4). If this arrangement is followed by an inversion of t s o p, (1) results. Finally, an inversion of r q p o will yield (2).

18. Problems occur during meiosis, not mitosis. Problems are worse in the odd-ploid organisms such as triploids because of the production of aneuploid gametes (see fig. 8.28).

20. Odd-ploid autopolyploids have meiotic problems, producing aneuploid gametes, as do allopolyploids, because of a lack of pairing partners. Amphidiploids should have few or no meiotic problems.

22. XX, 0, and X (unaffected) eggs will result.

24. 32. The gamete from P will have 9 chromosomes, and the gamete from U will have 7 chromosomes. The original zygote will have 16 chromosomes, but none of these will pair. To be fertile, each chromosome must be duplicated to yield 32 chromosomes.

26. Either. The woman is X^CX^c, the man is X^cY, and the child is X^cx^cY. If meiosis is normal in the woman, the egg gets X^c, and the sperm must contain X^cY. If meiosis is normal in the male, the sperm gets Y. If the chromatids of the X^c chromosome fail to separate during the second division in the female, the egg gets X^cX^c.

28. Since this karyotype is a tetraploid, it is likely that after fertilization all the chromosomes duplicated.

Critical Thinking Question

2. If the plants were closely enough related, both in physiology and chromosome number, it is conceivable that a successful hybrid could be created. Perhaps chromosome doubling would be necessary to achieve successful meiosis. Although the odds of this are low, the odds are even lower that the hybrid would combine the desired attributes of each species. Similar experiments done in the past have rarely been successful even when viable hybrids were produced.

CHAPTER 9
CHEMISTRY OF THE GENE

Key Genetic Concepts

The Genetic Material: As you learned in your introductory biology course and in chapter one, deoxyribonucleic acid, or DNA, is the genetic material in most organisms. Studies by Griffith, Avery et al., and Hershey and Chase provided the evidence necessary to identify DNA as the genetic material. This molecule directs, totally or in part, all of the functions and characteristics of an organism. In order to serve as the genetic material, a molecule such as DNA must possess several characteristics. First, DNA must be able to control the production of enzymes responsible for cell growth, development, and functioning. Second, the genetic material must be capable of self replication. Finally, the genetic material must be able to accommodate change (**mutation**).

Structure and Chemistry of DNA and RNA: DNA and RNA are nucleic acids made up of **nucleotide** subunits. Each nucleotide is made up of a sugar (**ribose** in RNA and **deoxyribose** in DNA), a phosphate, and a nitrogenous base. Nucleotides are linked together by a **phosphodiester bond** between the 5′ phosphate of one nucleotide and the 3′ hydroxyl of the adjacent nucleotide. X-ray crystallography indicates that the structure of DNA is a helix. Both DNA and RNA contain the **purine** bases adenine and guanine. DNA contains the **pyrimidines** cytosine and thymine while RNA contains the pyrimidines cytosine and uracil. In DNA, the percentage of adenine equals the percentage of thymine and the amount of cytosine equals the amount of guanine (**Chargaff's rules**) because adenine pairs with thymine and cytosine with guanine. Two hydrogen bonds form between adenine and thymine while three bonds connect guanine and cytosine. Watson and **Crick** proposed a model structure for DNA. They described the DNA molecule as a double helix or twisted ladder with the sides of the ladder (sugar-phosphate backbones) running in opposite directions (**antiparallel**). The A=T and C≡G base pairs make up the rungs of the ladder.

Prokaryotic DNA Replication: Each strand of the DNA molecule serves as a template for the synthesis of a new DNA strand. Replication is **semiconservative**: each strand is conserved in replication. The DNA helix is unwound by the enzyme **helicase** and the single strands of DNA are stabilized by Ssb proteins. DNA replication begins at the Y-junction and is an enzymatic process. Each nucleotide is added (polymerized) into a growing strand by a DNA polymerase. **DNA polymerase I** fills in small DNA segments during replication and repair while **DNA polymerase II** acts as a repair enzyme. **DNA polymerase III** is the primary enzyme involved in replication: it catalyzes a bond between the 3′-OH carbon of the last nucleotide in the strand and the 5′-PO_4 carbon of the new nucleotide. Because the strands of DNA are antiparallel and the enzyme polymerizes DNA in one direction only, each strand is replicated differently.

- Replication occurs continuously using the 3′ to 5′ template strand. A free 3′-OH in the growing strand is necessary for polymerization to occur because the enzyme adds bases in only the 5′ to 3′ direction. The newly synthesized strand is called the **leading strand**.
- DNA synthesis using the complementary strand is discontinuous. The **lagging strand** is synthesized in pieces (**Okazaki fragments**) away from the Y-junction. **Primase** first adds stretchs of RNA (**primers**) between which Okazaki fragments composed of DNA are later synthesized. The primer provides a free 3′-OH group needed by DNA

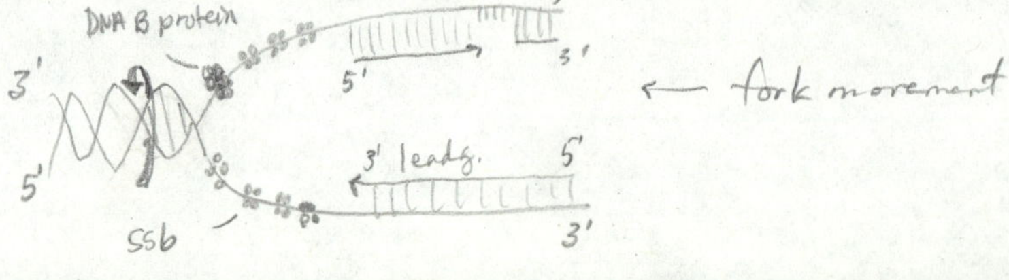

polymerase III to elongate the Okazaki fragments. DNA polymerase adds nucleotides into the new strand until another primer is encountered. Upon completion of all fragments, DNA polymerase I removes the RNA primers and replaces them with DNA. Finally, DNA **ligase** joins the Okazaki fragments by means of phosphodiester bonds.

Eukaryotic DNA Replication: In eukaryotes, there are usually multiple origins of replication and additional polymerase enzymes (see Tamarin Table 9.5).

Problem-Solving Hints

Hint 1: RNA can be identified by the presence of uracil and absence of thymine. DNA contains thymine but not uracil.

Hint 2: Chargaff's rules provide a means of determining the base composition of the DNA molecule. Knowledge of the amount of one base can be used to determine the percentage of the remaining bases in the molecule. For example, 10% adenine indicates that the molecule contains 10% thymine. The remaining 80% of the bases are guanine and cytosine, in approximately equal proportions (40% C and 40% G).

Hint 3: Chargaff's rules can also be used to determine if a molecule is single stranded or double stranded. The amount of adenine equals the amount of thymine and the amount of cytosine equals the amount of guanine only in a double stranded DNA molecule.

Hint 4: The temperature at which DNA denatures, or separates into single strands, provides insights to the base composition of the molecule. The three hydrogen bonds between cytosine and guanine are more difficult to break than the two bonds between adenine and thymine. Therefore, the higher the temperature needed to denature DNA, the higher the guanine-cytosine composition.

Hint 5: DNA polymerase III requires a free 3′-OH in the growing strand for polymerization of a new nucleotide. Continuous replication can only occur in the 5′ to 3′ direction using the 3′ to 5′ template strand.

Practice Problems

Problem 9-1. Modern molecular genetics is based upon the assumption that DNA possesses the characteristics listed below. Provide a sample drawing and brief explanation to illustrate how DNA fulfills assumption (a) below.

a. ability to replicate itself

b. ability to store genetic information specifying the form and function of organism

c. ability to express the stored information

d. ability to accommodate changes is the structure of the molecule (mutation)

Problem 9-2. Examine the following DNA molecule in the process of replication. Identify A - M below.

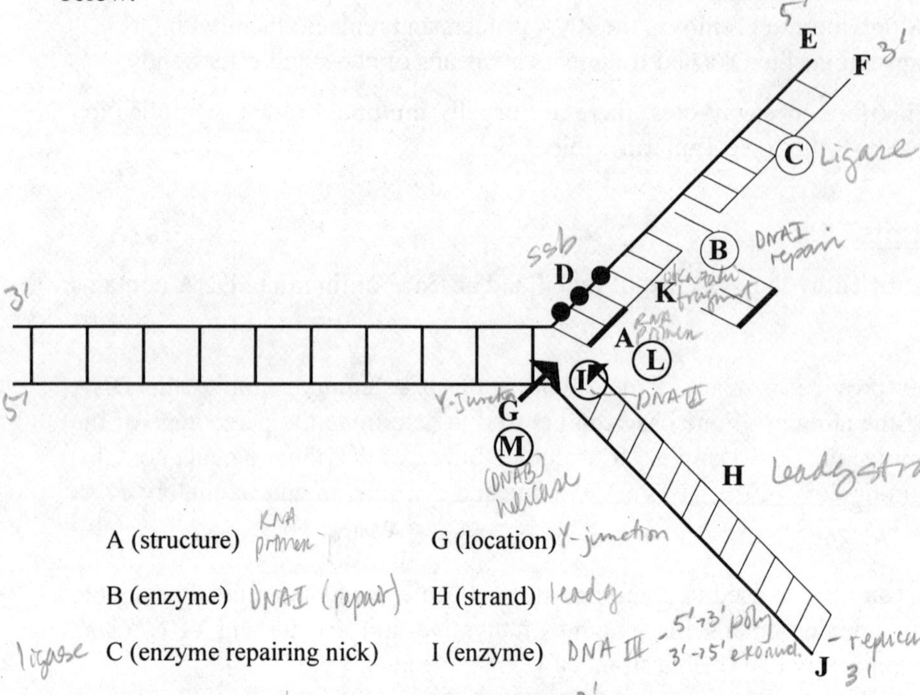

A (structure)

B (enzyme)

C (enzyme repairing nick)

D (proteins)

E (polarity)

F (polarity)

G (location)

H (strand)

I (enzyme)

J (polarity)

K (structure)

L (enzyme inserting segment A)

M (enzyme creating location G)

Problem 9-3. Examine the following generalized representation of a strand of nucleic acid. Determine if the strand is DNA or RNA. Identify the 5′ and 3′ ends of the molecule.

End "A"

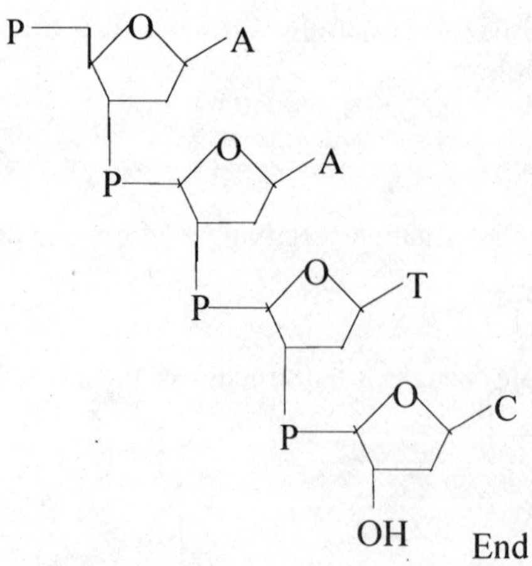

End "B"

Problem 9-4. Chargaff's ratios were instrumental in the development of the Watson-Crick model of DNA structure. Assume that Chargaff's data were as follows: 35% A; 20% T; 20% C; and 35% G. Assume further that the data of Wilkins and Franklin were unchanged. What conclusions, if any, would have been made by Watson and Crick?

Problem 9-5. Consider the following DNA base compositions: (A+G)/(T+C) = 1.0 and (A+T)/(C+G) = 1.5. Is the molecule single or double stranded?

Problem 9-6. Provide the strand that is complementary to the one below.

3′ TAAGCTTAACGCGCATCAGCGCGCTCCGATCGATCGATCAGCGATCAGAGGGAAG 5′

Problem 9-7. Would the above strand serve as the template for the leading strand or the lagging strand (assuming that the Y-junction is moving from right to left)? If it is the template for the lagging strand, what bases would primase insert if the first 10 bases initiated an Okazaki fragment?

Problem 9-8. How would the results of Griffith, Avery et al., and Hershey-Chase have differed if the genetic material was RNA rather than DNA? If the genetic material was protein rather than DNA?

Problem 9-9. Consider each of the problems (a-e) associated with prokaryotic DNA replication. For each, determine which enzyme(s) is (are) defective?

a. DNA synthesis is never completed because Okazaki fragments are not joined.

b. DNA synthesis of the lagging strand is never initiated.

c. Replication is never initiated although the DNA is unwound.

d. The newly synthesized DNA contains mismatched pairs.

e. New DNA strand contains some RNA bases

Problem 9-10. Examine the base composition of each DNA molecule below. Order the molecules (only one strand is shown) according to the relative temperature required to denature the molecule (highest to lowest).

a. GCTATAATAC

b. GGCTATCGGG

c. TTATCTAATT

General Questions

Question 9-1. You are provided with three samples of DNA and asked to determine which of the three has the highest G-C ratio. How can you quickly make this assessment?

Question 9-2. What evidence indicates that adenine pairs with thymine and cytosine pairs with guanine in the DNA double helix?

Solutions to Practice Problems

P9-1. DNA replication is a semi-conservative process: one strand of the double helix is preserved during each replication. Each strand serves as a template for the synthesis of a new strand

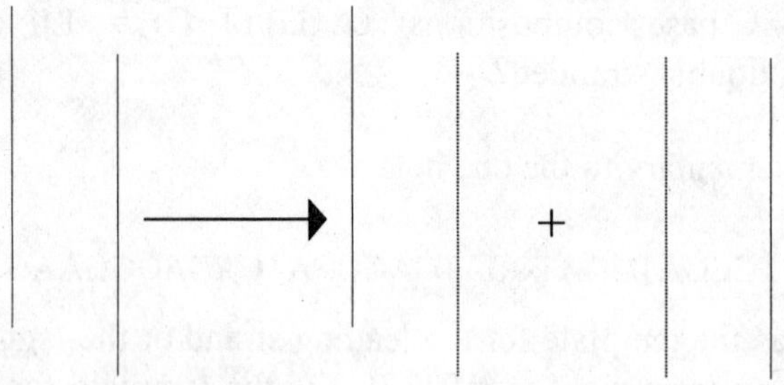

P9-2.

A RNA primer

B DNA polymerase I

C Ligase

D Ssb proteins

E 5′ end

F 3′ end

G Y-junction

H Leading strand

I DNA polymerase III

J 3′ end

K Okazaki fragment

L Primase

M Helicase

P9-3. This molecule contains thymine (T) and is therefore composed of DNA. If the molecule were RNA, the base uracil (U) would have been present (*Hint 1*). End "A" contains a phosphate group indicating that this is the 5′ end of the molecule. End "B" contains an exposed -OH group indicating that it is the 3′ end of the molecule.

P9-4. The data provided indicate that the bases pair in the following way: A-G and T-C. In this case, the two pyrimidine bases pair and the two purine bases pair. This data would still indicate that the molecule is double stranded. However, this base pairing would provide the incorrect thickness for the molecule based on the data of Wilkins and Franklin.

P9-5. In a double stranded DNA molecule, (A+G) = (T+C) because of the specific base pairings in DNA. In this case, (A+G)/(T+C) = 1.0. The data given are consistent with a double stranded DNA molecule. This molecule has a high A-T content because (A+T)/(C+G) exceed 1.0.

P9-6.

5′ ATTCGAATTGCGCGTAGTCGCGCGAGGCTAGCTAGCTAGTCGCTAGTCTCCCTTC 3′

P9-7. The DNA strand in the above problem would serve as the template for the lagging strand, assuming that the 3′ end of the molecule is the Y-junction (*Hint 5*). Following the rules of complementarity, the RNA primer sequence would be 5′ AUUCGAAUUG 3′.

P9-8. Regardless of the nature of the genetic material, the results of Griffith would not have differed. Griffith identified a "transforming" substance but did not identify which material was responsible for the transformation. In contrast, the results of Avery et al. would have differed significantly if DNA were not the genetic material. For example, if RNA had been the genetic material, the cell fraction containing RNA would have transformed bacterial cells. Likewise, the fraction containing proteins would have been biologically active if the genetic material were protein. The results of the Hershey-Chase experiments would have been identical if the genetic material were RNA because both contain phosphorus. The results would have differed if the material were protein. In this case, the progeny phage resulting from a cell provided with sodium labeled medium would be strongly labeled.

P9-9. Each of the problems described below probably arose from a single defective enzyme.

a. Ligase

b. Primase

c. DNA polymerase III

d. DNA polymerase I or II

e. DNA polymerase I

P9-10. Molecule (b) has the highest G-C content and will denature at the highest temperature. Molecule (a) will denature at an intermediate temperature. Molecule (c) has the highest A-T content and will denature at the lowest temperature. Refer to question 9-1 answer below for an additional explanation concerning the relationship between base composition and temperature.

Answers to General Questions

Q9-1. At elevated temperatures, DNA denatures. The relative temperature at which a DNA molecule denatures is related to its specific base composition (*Hint 4*). There are three hydrogen bonds joining cytosine and guanine while only two hydrogen bonds form between adenine and thymine. Three hydrogen bonds require more energy to break than two hydrogen bonds. Therefore, G-C pairs denature at a higher temperature than A-T bonds. Determine at what temperature the three sample denature to determine which has the highest G-C content.

Q9-2. The specific pairing of bases in DNA is supported by two pieces of evidence: Chargaff's ratios and X-ray crystallography data. Chargaff found that the ratio of purines to pyrimidines was equal and that the amount of adenine equales that of thymine while the amount of guanine equals that of cytosine. This evidence suggests that the molecule is double stranded with specific paring between adenine and thymine, guanine and cytosine. X-ray crystallography also suggested that the structure was a double helix. The measurements made correspond with the pairing of a purine and a pyrimidine.

Answers to Even Numbered End of Chapter Exercises, Problems and Critical Thinking Questions

2. Only proteins and nucleic acids were ever considered seriously.

4. See figure *9.13b*.

6. a. double-stranded DNA; b. single-stranded DNA; c. double-stranded RNA; d. single-stranded DNA; e. single-stranded RNA; f. DNA-RNA hybrid. First, look at what bases are present. If T is present, the molecule is DNA, and if U is present, the molecule is RNA. Remember that for a double-stranded molecule A must equal T (U) and G must equal C. For molecule (a) we see that A = T and G = C. In molecule (b), purines (A + G) < pyrimidines (C + T). In double-stranded molecules, purines = pyrimidines. Molecule (c) is consistent with double-stranded RNA. In (d) and (e), G is not equal to C,

therefore, the molecules are single-stranded. In (f), T and U are present, such that T + U = A. A DNA-RNA hybrid has U opposite A in RNA and T opposite A in DNA: thus A = T + U.

8. (lowest) 69°C, 73°C, 78°C, 82°C, 84°C (highest). Remember that a G-C pair has three hydrogen bonds and thus requires more energy to be broken than an A-T pair. Therefore, the higher the melting temperature, the higher the G-C content. Simply arrange the molecules from lowest to highest melting temperature.

10. A = 13.3%, G = 26.7%, C = 15%, T = 45%. Begin by writing the given information as equations:

$$G = 2A;\ T = 3C;\ (C + T)/(A + G) = 1.5$$

Now substitute for G and T in the third equation:

$$(C + 3C)/(A + 2A) = 1.5$$

$$4C/3A = 1.5$$

$$4C = 4.5A$$

$$C = \frac{4.5A}{4}$$

No, remember that the sum of all bases must = 100%.

$$A + G + C + T = 100$$

Substitute equalities into this equation:

$$A + 2A + \frac{4.5A}{4} + \frac{13.5A}{4} = 1$$

$$3A + 4.5A = 1$$

$$7.5A = 1$$

$$A = \frac{1}{7.5} = 0.133 = 13.3\%$$

$$G = 2A = 2(13.3) = 26.7\%$$

$$C = \frac{(4.5)(0.133)}{4} = 15\%$$

$$T = 3C = 45\%$$

12.

Conservative

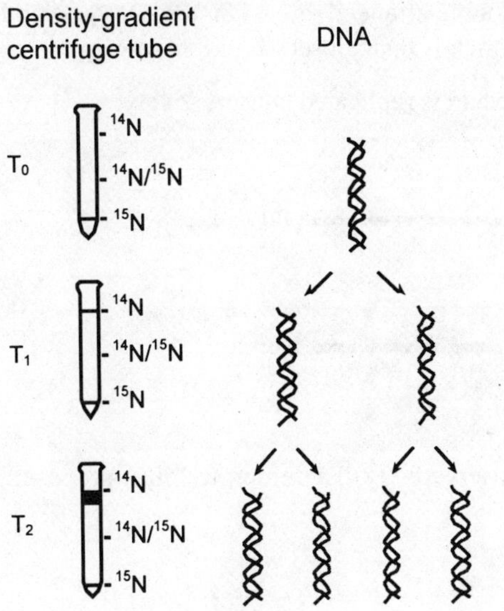

Dispersive

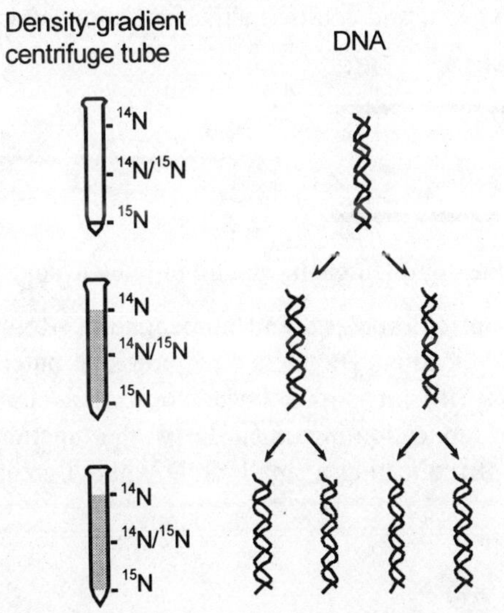

14.

Original	5′-ATTCTTGGCATTCGC-3′
DNA complement	3′-TAAGAACCGTAAGCG-5′
RNA complement	3′-UAAGAACCGUAAGCG-5′

Replication is 5′→3′.

16. Continuous replication occurs 5′→3′creating a leading strand; discontinuous replication occurs 5′→3′ backward from the replicating fork on the 5′→3′ template creating a lagging strand. Discontinuous replication is necessitated because none of the DNA polymerases acts in the 3′→5′ direction.

18. See figure 9.31 for the origin, figure 9.34 for the continuation, and figure 9.37 for the termination of DNA replication in *E. coli*.

20. Possible mechanism of DNA gyrase involves a double-stranded break in the DNA. Another double-stranded helix is then passed through the break, which is then closed (figure following).

22. You could imagine a situation in which a single molecule is replicated conservatively.

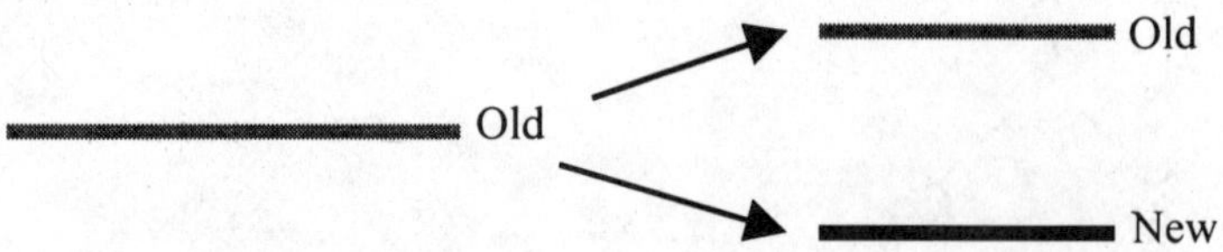

Unfortunately, all nucleic acids seem to replicate semiconservatively. Therefore, we must propose that a complementary strand (–) is made from the first:

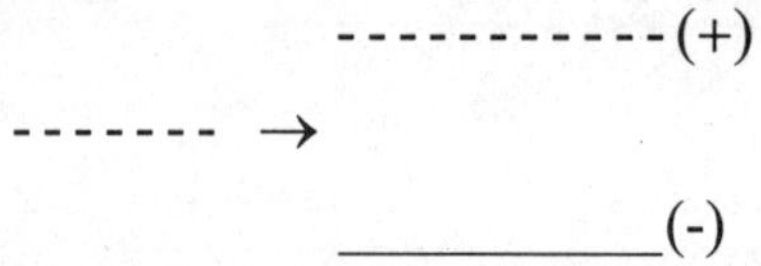

We now need to make more (+) strands from the (–). The (+) strand could separate and allow new (+) strands to be made and peeled off:

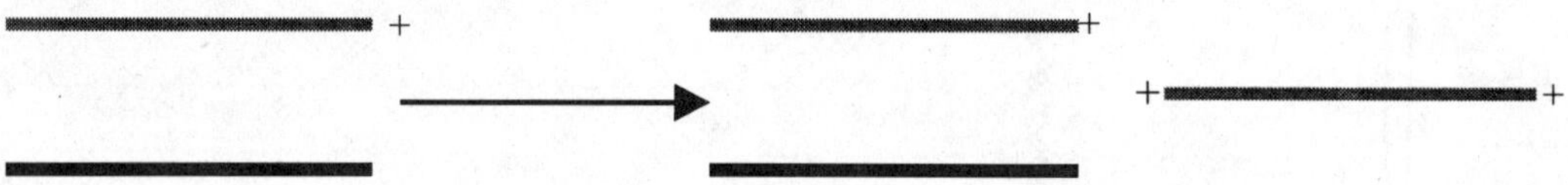

Other possibilities also exist as long as you keep in mind the complementary nature of nucleic acids.

24. Theta structure: prokaryotic DNA replication. D-loop: chloroplasts and mitochondria. Rolling-circle: bacterial conjugation. Bubbles: eukaryotic DNA replication. The theta structure is an outcome of replicating a closed circle; the rolling-circle is a way to transport a linear copy of a circular chromosome; bubbles are outcomes of multiple origins of replication in a linear DNA; the function of D-loops is not obvious. It may be a more efficient way of replicating a small circle without creating Okazaki fragments.

Critical Thinking Question

2. In their 1953 paper, Watson and Crick give one reason why DNA has the deoxyribose sugar: "It is probably impossible to build this structure with a ribose sugar in place of the deoxyribose, as the extra oxygen atom would make too close a van der Waals contact." (See Box 9.1.) From an evolutionary standpoint, it is also useful for the cell to be able to differentiate RNA from DNA for the removal of RNA primers during DNA replication; the oxygen difference in the sugars would provide that ability. As for the difference in pyrimidines (thymine vs. uracil), we believe the difference has to do with the cell's ability to repair spontaneously mutated bases. Although we will go into this in more detail in chapter 16, we can mention here that the most common spontaneous mutation of bases is of cytosine to uracil by deamination (see fig. 9.8). If uracil were being created spontaneously in DNA and uracil were a normal base in DNA, then the repair systems would not be able to differentiate a normal uracil from a mutated uracil and many mutations would go unrepaired. Hence early in evolution, thymine was probably substituted for uracil in DNA to alleviate that problem.

CHAPTER 10
GENE EXPRESSION: TRANSCRIPTION

Key Genetic Concepts

Transcription versus Translation: Transcription involves the transfer of information stored in DNA to an RNA molecule. The information encoded in the RNA molecule is subsequently used to assemble a specific protein in the process of **translation**.

The Function of RNA: DNA directs the formation of proteins, especially enzymes. However, DNA is located in the cell nucleus (in eukaryotes) and protein synthesis occurs in association with the ribosomes in the cytoplasm. If DNA does not leave the nucleus, how can it direct protein synthesis? The answer: **messenger RNA (mRNA)** carries the sequence information of DNA to the ribosomes. Other important RNA molecules are **transfer RNA (tRNA)**, which brings the amino acids to the ribosome, and ribosomal **RNA (rRNA)** which is a structural component of the ribosome. Transfer RNA is essential to the process of protein synthesis because it ensures that the amino acids are inserted in the proper sequence. This is accomplished by complementary base pairing between the sequence of three bases on the tRNA molecule (the **anticodon**) and the three base sequence on the mRNA (the **codon**) which specifies the amino acid to be inserted

Prokaryotic Transcription: The **holoenzyme RNA polymerase** synthesizes mRNA using a DNA template in the process of transcription. The core enzyme component of RNA polymerase carries out the actual polymerization of bases while the **sigma factor** recognizes signals which begin the process. The enzyme associates on the region of the DNA molecule described as the **promoter**. Although not transcribed, the promoter is required to "position" the RNA polymerase for downstream transcription. A six base sequence (5′ TATAAT 3′ or a variation of this sequence), known as the **Pribnow box**, is recognized within the promoter. Further upstream (-35) of transcription is a second recognized sequence (5′ TTGACA 3′). Once initiated, RNA polymerase begins adding nucleotides to the growing strand of mRNA in the 5′ to 3′ direction. Specifically, the enzyme adds cytosine when guanine is encountered on the DNA template, guanine when cytosine is encountered, adenine when thymine occurs in the template, and uracil to correspond to adenine. In this way, the messenger RNA has the opposite sequence of the DNA template and the same sequence as the **coding strand**. Synthesis of mRNA continues until a **terminator** is encountered (see Tamarin Figure 10.10).

Eukaryotic Transcription: Transcription in eukaryotes differs from that of prokaryotes in three major ways. First, because the DNA is contained in a nucleus (eukaryotes), transcription and protein synthesis do not occur simultaneously as they do in prokaryotes. Second, there are three active RNA polymerase enzymes in eukaryotes. RNA polymerase I transcribes the nuclear organizer, RNA polymerase II transcribes most genes, and RNA polymerase III transcribes rRNA and tRNA genes. The promoter for all three polymerases is a sequence (5′ TATAAA 3′) referred to as a TATA box. Third, eukaryotic mRNA is modified posttranscriptionally while prokaryotic mRNA is not. Modifications include the addition of a 5′ cap to facilitate recognition at the ribosome and the addition of a 3′ poly-A for stability or transportation of the molecule out of the nucleus. Additional sequences, called **intervening sequences** or (**introns**), are transcribed but are never translated into protein are also removed posttranscriptionally.

The Central Dogma: The **central dogma** focuses on information transfer at the molecular level. DNA produces identical copies of itself during replication (Chapter 9) as does RNA. Information flows from the base sequence of DNA to an RNA molecule (transcription). The information encoded in the RNA molecule is then used to assemble proteins (translation). In some viruses, information encoded in RNA can be used to synthesize DNA in a process known as **reverse transcription**. DNA does not directly serve as the template for protein synthesis in nature. Information transfer from a protein to another molecule is apparently forbidden.

Problem-Solving Hints

Hint 1: The processes of transcription and translation can be easily distinguished by examining which molecules are involved in the information transfer. If a DNA template is used to synthesize an RNA molecule, transcription has occurred. If an RNA template directs the formation of an amino acid sequence, the process is translation.

Hint 2: Messenger RNA is synthesized in the 5′ to 3′ direction using a 3′ to 5′ template DNA strand. Although either strand can serve as the template, only one strand is transcribed. Transcription of both strands may result in the formation of double stranded mRNA molecule which cannot be translated.

Hint 3: The base sequences given for the promoter region (above) are on the coding strand. The complementary strand is actually transcribed.

Hint 4: The coding strand of the DNA molecule has the same base sequence as the messenger RNA molecule (with U substituted for T in RNA) transcribed from the template strand.

Hint 5: Post-transcriptional modification of mRNA occurs before *eukaryotic* translation.

Practice Problems

Problem 10-1. A mRNA molecule is isolated and determined to be composed of 30% A, 33% U, 18% G, and 19% C. Using this information, determine the base composition of the DNA template and the DNA coding strand.

Problem 10-2. Consider the DNA molecule below. What mRNA would be synthesized if the top strand served as the template? What would be the mRNA sequence synthesized from the bottom strand?

3′ GGCCATTCTACCAAAATCGC 5′

5′ CCGGTAAGAAGGTTTTAGCG 3′

Problem 10-3. If both strands of the molecule in problem 2 are transcribed, what is the likely consequence? Why is only one or the other strand of DNA typically transcribed?

Problem 10-4. Consider the DNA molecule below. Identify the Pribnow box, the direction of transcription, the beginning of the gene, the mRNA sequence, the polarity of the mRNA molecule, the upstream direction, the downstream direction, the coding strand, and the template strand. The base indicated in bold is the first to be transcribed.

5′ TAGGACCGCTATAATGCAGCAGAA**A**GCGATGGCCACCCCACCAAAATCGCCTCCACGC 3′

3′ ATCCT GGCGATATTACGTCGTCTT**T**CGCTACCCGGTGGGGTGGTTTTAGCGGAGGTGCG 5′

Problem 10-5. Given below are the sequences found in a DNA template (top) and the final mRNA product found at the ribosome (bottom). How can you explain the differences in the base pair composition of these two molecules?

3′ TACCGGCCTATCTAGGCCATCGCATACGGCAGGTTTTACGCGTACGCCTTTTAATTA 5′

5′ AUGGCCGGGGUAGCGUAUGCCAAUGCGCAUGCGGAAAAAAU 3′

General Questions

Question 10-1. Compare the consequences of inserting an incorrect base in the DNA template during replication and inserting an incorrect base in the mRNA during transcription.

Solutions to Practice Problems

P10-1. The mRNA has the same sequence and composition (T substituted for U) as the coding strand of DNA (*Hint 4*). Therefore, you would expect to find 30% A, 33% T, 18% G, and 19% C in the coding strand. The template strand has the opposite sequence as the mRNA. Because A and T pair, as well as C and G, you would find 30% T, 33% A, 18% C, and 19% G.

P10-2. Remember that RNA polymerase synthesizes mRNA in the 5′ to 3′ direction using 3′ to 5′ template (*Hint 2*). The top strand in this example will be transcribed from left to right yielding the mRNA: 5′ CCGGUAAGAUGGUUUUAGCG 3′. The bottom strand will be transcribed from right to left yielding the mRNA: 5′ CGCUAAAACCUUCUUACCGG 3′.

P10-3. Notice that the mRNA sequences are complementary to each other (5′ to 3′ mRNA pairing with the 3′ to 5′ mRNA). The two mRNAs would base pair forming a double stranded RNA molecule which could not be translated.

P10-4. Remember that the Pribnow box sequence (TATAAT) occurs on the coding strand although the 3′ to 5′ strand is transcribed (*Hint 3*). The top strand is the coding strand while the bottom strand is the template strand. The Pribnow box is in bold. The initiation codon specifying the beginning of the gene is indicated in bold italic.

5′ TAGGACCGC**TATAAT**GCAGCAGAA**A**GCGATGGGCCACCCCACCAAAATCGCCTCCACGC 3′
3′ ATCCT GGCG**ATATTA**CGTCGTCTT**T**CGC***TAC***CCGGTGGGGTGGTTTTAGCGGAGGTGCG 5′

direction of transcription →

← upstream | downstream →

mRNA: 5′ AGCGAUGGGCCAUUCUACCAAAAUCGCCUCCACGC 3′

P10-5. The mRNA has been modified posttranscriptionally (*Hint 5*). Introns (bold sequences) have been removed from the originally transcribed mRNA.

3′ TACCGGCC**TATCTAGG**CCATCGCATACGG**CAGGTT**TTACGCGTACGCCT T T TT**AAT**TA 5′

5′ AUGGCCGG——— GGUAGCGUAUGCC ——— AAUGCGCAUGCGGAAAAA ——AU3′

Answers to General Questions

Q10-1. Inserting an incorrect base during replication will create an error that is passed on to subsequent cells. Changes in the DNA sequence will persist in subsequent generations of cells. Inserting an incorrect base during transcription will potentially result in the formation of an abnormal protein. However, this mistake will not persist, and will not be transmitted, if the DNA template is unmodified.

Answers to Even Numbered End of Chapter Exercises, Problems and Critical Thinking Questions

2. Hybridization of nucleic acids is carried out between species to determine the extent to which it occurs. Presumably, the greater the amount of hybridization, the more similar the genomes of two species and therefore the more closely they are related in an evolutionary sense (see chapter 21).

4. In prokaryotes, the transcription start signal is the promoter with two or three recognition sequences, recognized with the aid of the sigma factor of RNA polymerase. Transcription stop signals are terminators, some of which require rho factors. Both rho-dependent and rho-independent terminators cause stem-loops to form in the RNA, pausing the polymerase. In eukaryotes, the core promoter has the TATA sequence recognized by TFIID. Terminators have pause and possibly release sequences, probably recognized by specific proteins. Polycistronic transcripts are routine in prokaryotes, rare in eukaryotes.

6. With no sigma factors, there would be random starts of transcription; with no rho factor, there would be a failure to properly terminate transcription at rho-dependent termination sites.

8. Although the overall process is similar in prokaryotes and eukaryotes, differences exist in the number of different RNA polymerases and transcription factors in eukaryotes; exact DNA sequences recognized; polycistronic nature of prokaryotic DNA; and posttranscriptional processing in eukaryotes (splicing, capping, polyadenylation).

10. TATAAT (Pribnow box): –10 sequence of prokaryotic promoters

TTGACA: –35 sequence in prokaryotic promoters

TATA (Hogness box): Recognition sequence of core of eukaryotic promoters

TACTAAC: Consensus sequence around lariat branch point in nuclear introns

12. Rho-dependent terminators require the rho protein for successful termination. Both terminator types produce stem-loop structures in the RNA. Rho-independent terminators have, in addition, a sequence of uracil-containing nucleotides after the inverted repeat (see fig. 10.10).

14.

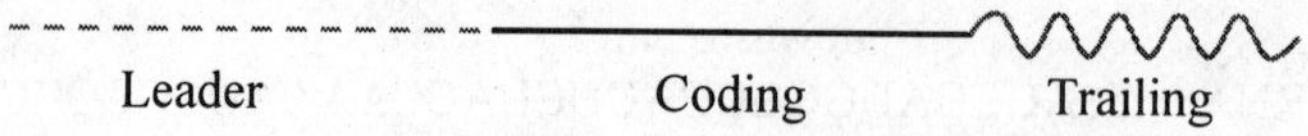

Prokaryotic

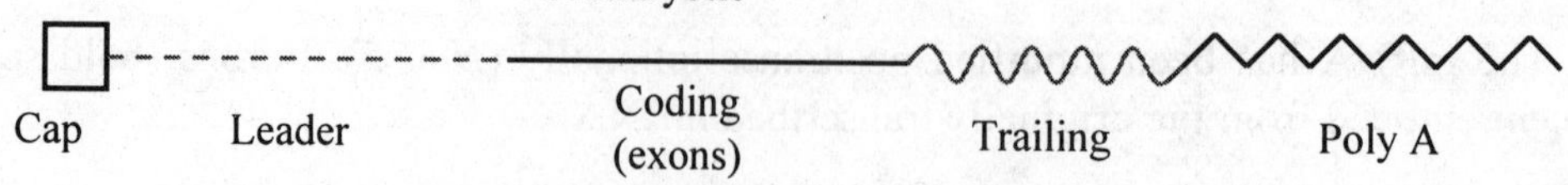

Eukaryotic

Both molecules will have 5′ and 3′ noncoding regions, the leader and trailing sequences, respectively. Eukaryotic messenger RNA has a 5′ cap and a 3′ poly-A tail.

16. Top 3′ ———— 5′

Bottom 5′ ——————— 3′

The top strand is transcribed (template strand).

RNA: 5′-G G G A U G C G G A A A G U C C A A-3′

Begin by writing the RNA sequence complementary to each DNA strand. Since the DNA fragment is from the beginning part of the gene, there must be a start signal, AUG, for protein synthesis. Transcription of the top strand yields such an RNA, but the bottom does not. Since transcription proceeds 5′ → 3′, the left end of the top strand must be the 3′ end. Remember that double-stranded nucleic acids are antiparallel.

18. 5′-A U A C C G U A C C A U G A A G G C C-3′

The Pribnow box is TAATGC, beginning eight bases from the 3′ end of the DNA. Transcription proceeds on the 3′ → 5′ strand producing a 5′ → 3′ RNA; therefore it moves left to right along the given DNA molecule because double-stranded nucleic acids are antiparallel.

20. All transcription could be rho independent, and we would see normal transcription at both temperatures. If transcription is rho dependent, long molecules could possibly be degraded rapidly, producing short molecules. Neither possibility is too likely since some genes seem to be rho independent and others rho dependent.

22. Introns do accumulate more mutations. Presumably, at most sites within an intron, mutations do not affect the phenotype and therefore are not selected against (see chapter 20).

24. Heterogeneous nuclear messenger RNAs are eukaryotic transcripts before posttranscriptional modifications. Small nuclear ribonucleoproteins are the components of the eukaryotic messenger RNA-splicing apparatus (spliceosome).

26. Recognition signals are shown in figures 10.30, 10.33, and 10.35.

28. See figures 10.30 and 10.31.

30. A transcriptional factor is a eukaryotic protein that is similar in function to prokaryotic sigma factors: they help the core polymerase recognize and attach to promoters. An enhancer is a eukaryotic DNA sequence that increases transcription when bound by an activator even if distant from the promoter itself.

32. The RNA is processed differently in the different tissues. In the pituitary, we see five introns and six coding regions, but in the adrenal, we see only four introns and five coding regions. The second DNA loop in the adrenal is larger than either the second or third loop in the pituitary. The third coding region must be removed in the adrenal, making the large second loop.

34. Prions are proteins that change the shape of proteins, not unlike many enzymes. The fact that they are mutant forms of a protein that change their unmutated equivalents into the mutated forms only makes them an oddity rather than a forbidden transfer in the central dogma. They do not represent a protein self-replication loop.

Critical Thinking Question

2. The intent of this question is to examine the way in which an activator could control transcription. Assume that the activator is needed to bind to its enhancer and from this to be in a position to bind to one of the proteins of the transcription complex (figure 10.24). With the activator bound, its interaction causes enhanced transcription; without it being bound, transcription is basal at best. Thus transcription of specific genes is dependent on the appropriate activators being present. We will see that this is the common mode of what we know about control of growth and development in eukaryotes.

CHAPTER 11
GENE EXPRESSION: TRANSLATION

Key Genetic Concepts

Transfer RNA. Transfer RNA molecules carry specific amino acids to the ribosome where they are incorporated into a growing polypeptide. The amino acid is bound to the 3′ end of the transfer RNA molecule by the enzyme **aminoacyl-tRNA synthetase**. A tRNA is said to be **charged** when the amino acid is attached. The tRNA carrying an amino acid is recognized by its **anticodon**. The anticodon ensures that the amino acids are inserted in the proper sequence: the anticodon pairs with the **codon** (sequence of three bases) within the mRNA.

Translation: Translation involves the assembly of a polypeptide or protein (sequence of amino acids) according to the specifications encoded in the mRNA molecule. Translation takes place at the ribosome and requires the presence of a mRNA, tRNAs carrying their specific amino acids, the 30S and 50S ribosomal subunits and associated rRNAs, and ATP. Translation begins when the two ribosomal subunits and the mRNA associate. The prokaryotic mRNA is recognized by a sequence of bases at the 5′ end of the mRNA referred to as the **Shine-Delgarno sequence**. This sequence base pairs in complementary fashion with the 16S rRNA. Two tRNA molecules are positioned within the ribosomal complex, the first is positioned within the **P site** and the second within the **A site**. The mRNA runs through the bottom of these sites (5′ to 3′).

The first amino acid to be added to a prokaryotic protein is N-formyl methionine which is specified by the initiation codon 5′ AUG 3′ found in the mRNA. The tRNA carrying methionine enters the P site of the ribosomal complex by recognition of its anticodon 3′ UAC 5′. The tRNA with an anticodon corresponding to the next codon in the mRNA enters the A site of the complex. The enzyme **peptidyl transferase** catalyzes a peptide bond between the first and second amino acids. The tRNA in the P site is now uncharged and drops out of the complex. The entire ribosomal complex shifts down the mRNA by one codon. The tRNA in the A site now shifts to the P site and the next tRNA molecule enters the A site. The existing dipeptide is bound to the next amino acid, the complex shifts, and the tRNA specified by the next codon enters. The process repeats until the entire polypeptide has been synthesized. The presence of a termination codon (**nonsense codons**) in the A site indicates the end of the encoded message. The sequences UAG, UAA, and UGA do not specify amino acids and act to terminate translation. Upon completion of the polypeptide, the ribosomal complex disassociates and the polypeptide folds into the final protein structure.

Prokaryotic versus Eukaryotic Translation: In prokaryotic organisms, the processes of transcription and translation are coupled. In eukaryotic organisms, the mRNA molecule must leave the nucleus where it is synthesized and move to the ribosomes where translation occurs. The prokaryotic mRNA may also contain information for the synthesis of several proteins (**polycistronic**). Eukaryotic mRNAs are almost always **monocistronic**. A complete comparison of prokaryotic and eukaryotic translation is given in Tamarin, Table 11.1.

The Genetic Code: The genetic code is composed of a triplet sequence (codon) in the mRNA molecule. The code is nonoverlapping, unpunctuated, unambiguous, degenerate, redundant, and nearly universal.

Problem-Solving Hints

Hint 1: The genetic code is found in the mRNA molecule. The table of codons found in Chapter 11 of Tamarin corresponds to mRNA codons, not tRNA anticodons or DNA bases.

Hint 2: The mRNA molecule is read from 5′ to 3′ and the polypeptide is synthesized from the amino terminal to the carboxyl terminal.

Hint 3: The anticodon of the tRNA base pairs with the mRNA codon ensuring that amino acids are inserted in the correct order.

Practice Problems

Problem 11-1. Modern molecular genetics is based upon the assumption that DNA possesses the characteristics listed below. Provide a sample drawing and brief explanation to illustrate how DNA fulfills assumptions (b) and (c) below.

a. ability to replicate itself

b. ability to store genetic information specifying the form and function of organism

c. ability to express the stored information

d. ability to accommodate changes is the structure of the molecule (mutation)

Problem 11-2. The sequence below is the DNA template sequence for two prokaryotic proteins. The entire sequence is transcribed as a single mRNA. What are the two gene products.

TCCTCCATACCGGGCGTTTGGGGGTATCTCCTCCATACCCGACGGGATTTTGGGCCGATTCCACT

Problem 11-3. How many amino acids comprise a protein whose mRNA is 304 bases in length? How long would an mRNA coding for a protein with 140 amino acids be? Take into consideration the length of the Shine-Delgarno sequence and the termination codon.

Problem 11-4. Complete the table below.

Coding DNA									
Template DNA									
mRNA	AUG								UAA
tRNA									
Amino Acids	Met	Gly	Cys	Lys	Tyr	Ala	Val	Ala	X

Problem 11-5. Given below are three fragments cleaved from the same gene. Identify the first, middle, and last fragment making up the gene.

DNA fragment 1: 5′ TTAAAAATACGG 3′

DNA fragment 2: 3′ TTTAAAGGCCGG 5′

DNA fragment 3: 5′ CCCCCGGCACAT 3′

Problem 11-6. For the mRNA sequence below, provide the resulting amino acid sequence given that the genetic code is (a) a nonoverlapping triplet; and (b) a triplet code overlapping by one base.

AUGCAGCAGGGCUCCGCACUGACCUAG

Problem 11-7. As we learned in this chapter, the genetic code is redundant: more than one codon can specify the same amino acid. If you wanted to synthesize a polypeptide from a mRNA sequence without redundancy, which amino acid(s) would you incorporate? Answer the same question for a mRNA with the maximum redundancy.

General Questions

Question 11-1. Organic matter is composed primarily of the elements C, H, O, N, and P. Which of these elements is found in DNA only, in protein only, or in both DNA and protein? Relate your answer to the design of the Hershey-Chase experiment described in Chapter 9.

Question 11-2. DNA can serve directly serve as the template for protein synthesis only in the laboratory. Why would this mechanism of information transfer be difficult in eukaryotic cells? The only "forbidden" pathways of information transfer are from protein to DNA or RNA. What difficulties would these mechanisms of transfer pose?

Question 11-3. Describe why puromycin, streptomycin, tetracycline, and chloramphenicol are effective antibiotics.

Solutions to Practice Problems

P11-1. Information is coded in the base sequence of the DNA. This information is transferred to mRNA during transcription. The mRNA contains the triplet genetic code (*Hint 1*). The information ultimately stored in the DNA molecule is expressed when the mRNA is translated at the ribosome.

P11-2. Each gene is separated by a Shine-Delgarno sequence (italic) and each has a start (bold) and stop codon (bold italic).

mRNA:
*AGGAGGU***AUG**GCCCGCAAACCCCCA***UAG****AGGAGGU***AUG**GGCUGCCCUAAAACC
CGGCUAAGG***UGA***

Amino Acid Sequence #1:
Met Ala Arg Lys Pro Pro

Amino Acid Sequence #2:

Met Gly Cys Pro Lys Thr Arg Leu Arg

P11-3. The Shine-Delgarno sequence is seven bases long and the stop codon is three bases long. Three base pairs equals one amino acid. An mRNA composed of 304 bases would produce a protein composed of (304 - 10) /3 or 98 amino acids. Likewise, a protein composed of 140 amino acids would be coded for by an mRNA with 140 x 3 (because the code is triplet) + 7 (the Shine-Dalgarno sequence) + 3(the stop codon) or a total of 430 bases.

P11-4. Notice that the third base can be A,T, C, or G in the case of Gly, Ala, and Val. Two possible bases can occupy the third base in the codons for Cys, Lys, and Tyr.

Coding DNA	ATG	GG?	TGT	AAA/G	TAT/C	GC?	GT?	GC?	TAA
Template DNA	TAC	CC?	ACA	TTT/C	ATA/G	CG?	CA?	CG?	ATT
mRNA	AUG	GG?	UGU/C	AAA/G	UAU/C	GC?	GU?	GC?	UAA
tRNA	UAC	CC?	ACA/G	UUU/C	AUA/G	CG?	CA?	CG?	AUU
Amino Acids	Met	Gly	Cys	Lys	Tyr	Ala	Val	Ala	X

P11-5. Recall that transcription produces a 5′ to 3′ mRNA using a 3′ to 5′ template. To identify the beginning of a gene, look for a TAC at the 3′ end (AUG in the mRNA is a start codon). To identify the last segment of the gene, look for a 3′ to 5′ sequence corresponding to a termination codon (UAG, UAA, UGA). Fragment three contains the beginning of the gene and fragment one contains the end of the gene. Fragment two contains the middle portion of the gene.

P11-6.

AUG	CAG	CAG	GGC	UCC	GCA	CUG	ACC	UAG
Met	Gln	Gln	Gly	Ser	Ala	Leu Thr		

versus

AUG		Met
GCA	Ala	
AGC		Ser
CAG		Gln
GGG		Gly
GCU		Ala
UCC	Ser	
CGC		Arg
CAC	His	
CUG		Leu
GAC	Asp	
CCU	Pro	
UAG	STOP	

P11-7. The two amino acids which are specified by only one codon are methionine (AUG) and tryptophan (UGG). The amino acids specified by the largest number of different codons are leucine (UUA, UUG, CUU, CUC, CUA, CUG), serine (UCU, UCC, UCA, UCG, AGU, AGC), and arginine (AGA, AGG, CGU, CGC, CGA, CGG).

Answers to General Questions

Q11-1. Both nucleic acids and proteins contain carbon, hydrogen, oxygen and nitrogen. Only DNA contains phosphorus while sulfur is found only in protein. This difference made the Hershey-Chase experiment possible. In this experiment, the researchers needed a reliable way to distinguish between DNA and protein. Radioactive sulfur was use to label protein while radioactive phosphorus made it possible to detect DNA.

Q11-2. In eukaryotic cells, DNA is contained within the nucleus while the site of protein synthesis is the ribosomes which are located in the cytoplasm. If DNA directly acted as the template for protein synthesis, the molecule would have to migrate from the nucleus to the ribosome. There are several possible problems. First, can the molecule pass through the nuclear membrane? Second, could the molecule be damaged by enzymes in the cytoplasm? Third, what would control the movement of the DNA molecule? Finally, how would regulation be accomplished?

Q11-3. Many antibiotics, including those listed, work by interfering with the process of bacterial gene expression. These compounds interact specifically with the prokaryotic cellular machinery. They are effective antibiotics because they specifically interfere with the processes of prokaryotic organisms. Because of the similarities between mitochondria and bacteria, these substances can, however, interfere with transcription or translation of the mitochondrial genes.

Answers to Even Numbered End of Chapter Exercises, Problems and Critical Thinking Questions

2. For example, if a site on the ribosome was a temporary docking region for the nascent polypeptide, only one transfer RNA need be present at a time.

4. In prokaryotes, the Shine-Dalgarno sequence is used for messenger RNA-ribosomal alignment. In eukaryotes, the scanning hypothesis describes the need for a 5′ cap and scanning behavior. Because a 5′ cap is needed for ribosomal alignment, only one polypeptide can be synthesized per messenger RNA (there is generally only one site of ribosomal attachment per eukaryotic messenger RNA). (There can be some variation due to alternative start codons, internal ribosome entry sites, and read-through.)

6. The three nonsense codons are 5′-UAA-3′, 5′-UAG-3′, and 5′-UGA-3′. They all begin with a pyrimidine (uracil) and have purines in the second and third positions. The theoretical anticodons are 3′-AUU/I-5′, 3′-AUC/U-5′, and 3′-ACU/I-5′, respectively.

8. An antibiotic would be effective if it interfered with any step of the prokaryotic translation process. Thus we could have antibiotics that interfered with any of the prokaryotic initiation, elongation, translocation, and termination factors, any of the aminoacyl-tRNA synthetases, and any of the exposed proteins of the ribosome.

10. 5′-UUA-3′ (leucine) → 5′-UAA-3′. The consequence is that the growing polypeptide will be terminated at an improper point, probably producing a nonfunctioning enzyme or protein.

12. There would be blockage of further protein synthesis because of the N-terminal formyl group that prevents a peptide bond. The growing peptide would be stopped at that point.

14. RF1 and RF2 recognize nonsense codons. Their eukaryotic equivalent is eRF.

16. Puromycin does not disrupt eukaryotic translation because it does not bind to the eukaryotic ribosome for reasons that have to do with the general conformation of the eukaryotic translation system.

18. 441 bases. There are 3 bases per amino acid; thus, to code for 146 amino acids we need $3 \times 146 = 438$ bases. We also must include a stop codon, so $438 + 3 = 441$. In reality the mRNA is much longer than 441 bases, with a 5′ cap, poly-A tail, and leader and trailer DNA.

20. Five bases are needed for codons composed only of two base types ($2^5 = 32$, a number large enough to include different codons for the twenty amino acids); if double-stranded, the proportion of C and G would be equal.

22. There would be implications to evolutionary theory both in general and in reference to the evolution of the genetic code, indicating a long and isolated evolutionary lineage of the organism in question. Extraterrestrial organisms probably would use a different code system. In other words, there is no obvious structural or functional relationship between a nucleotide base and an amino acid.

24. GGC. The first polymer produces repeated reading frames of GCG-CGC. Therefore, GCG and CGC represent alanine and arginine, but we cannot determine which is which. The second polymer produces repeated sequences of either CGG, GGC, or GCG. We know that GCG codes for either alanine or arginine. CGC and CGG probably code for the same amino acid (wobble or unmixed family), either alanine or arginine. The final combination, GGC, is not similar to the original codons and probably codes for glycine.

26. Six. Begin with CUC, and then list all possible codons in which the first C is changed; then list all possible codons in which only the U is changed. Finally, list all codons that result in a change of the second C.

List 1	*List 2*	*List 3*
AUC: *ile*	CAC: *his*	CUG: *leu*
UUC: *phe*	CGC: *arg*	CUA: *leu*
GUC: *val*	CCC: *pro*	CUU: *leu*

28. Not apparent. Either glutamate codon (GAA or GAG) can mutate in one step to valine (GUX) or lysine (AAA or AAG).

30. GGX GCU AGC CAU UGC CUC UUC/U. Mutant 1 is shorter than normal, indicating the cysteine codon mutated to a stop codon. Since cysteine is UGU/C, the stop must be UGA in the mutant. In mutant 2, histidine is replaced by leucine; CAU/C $\rightarrow$ CUX. Changing the A to a U will give a leucine codon. Only the third mutant allows us to determine the normal sequence. All amino acids after the first are changed, indicating a frame shift, an addition or deletion of one base. We must therefore line up the two possible sequences. Write down the possible sequences for the normal:

GGX GCX UCX　　CAU/C UGU/C UUA/G UUC/U

AGU/C　　CUX

Then write down the possible sequences for mutant 3:

GGX GUX GCX AUU/C/A GUX UCX

AGU/C

By aligning these sequences, the change can be seen as a deletion of the fifth base in the normal sequence (C). Therefore, only certain of the original possibilities can exist.

Critical Thinking Question

2. Since the E site exists, we must look at the process of translocation to see what function it serves. Presumably, at this point in our understanding, the E site is a necessity because of the way that EF-G translocates the ribosome. That is, EF-G moves the ribosome in relation to the messenger and transfer RNAs but does not simultaneously eject the uncharged transfer RNA in the P site. Thus, a site is needed on the ribosome to hold that transfer RNA until after translocation when it can be ejected. The E site is not a conceptual necessity and was not a part of many early models of the ribosome and translocation.

CHAPTER 12
RECOMBINANT DNA TECHNOLOGY

Key Genetic Concepts

Restriction endonucleases occur naturally in bacteria where they cleave foreign DNA. This ability makes them invaluable in recombinant DNA protocols. Primarily used are type II endonucleases which cleave the DNA at a known and predictable sequence called a **restriction site**. DNA molecules that have been cleaved can be rejoined or joined with another molecule possessing the same free ends.

Vectors Carrying Foreign DNA: Plasmids, which were first introduced in Chapter 7, can be used to transfer pieces of DNA to prokaryotic host cells. A restriction enzyme can be used to cut both the target DNA and a plasmid (one cleavage site per plasmid), creating complementary sequences on the free ends of each. The two are mixed together with ligase which will join the free ends of the target DNA and the plasmid. Plasmids that have incorporated the target DNA (hybrids) are then used to transfer the target DNA to another cell. **Viruses** can also be used as vectors to carry small pieces of foreign DNA. **Charon phages** are the most common vehicles employed in recombinant DNA protocols. Bacterial plasmids have also been altered to contain a yeast centromere and origin of DNA replication which allows them to be replicated inside of the yeast host. These modified plasmids are called **yeast artificial chromosomes (YAC's)**. YAC's have allowed yeast cells to be utilized in recombinant DNA research thereby giving scientists an important eukaryotic model organism. In higher animals, SV40, a DNA tumor virus, is commonly used as a vector. In plants, a soil bacterium is used to introduce foreign genes.

Isolating Genes of Interest: Purified double-stranded DNA is obtained by (a) enzymatically converting mRNA to DNA (resulting in the production of **complementary DNA** or **cDNA**), (b) synthesizing a DNA molecule *in vitro*, or (c) cutting the entire genome of an organism into small pieces from which the desired gene is isolated.

- **Southern blotting** and hybridization are used to isolate the desired gene if the entire genome is digested with restriction enzymes. In this method, the fragments produced from the restriction enzyme digest are separated according to size by **gel electrophoresis**. While still on the gel, the separated DNA is transferred and permanently bound to a membrane (Southern blotting). This membrane is then immersed in a solution containing a radioactively labeled nucleic acid (a **probe**) possessing a base sequence complementary to one found in the target DNA. The probe will base pair with its complementary sequence. When exposed to photographic film, the gene of interest will be "highlighted" (**hybridization**). Hybridization can also be used to identify genes without an electrophoretic-separation step in a process know as **dot blotting**.
- **Western blotting** is a second method used to locate specific genes. In this method, antibodies specific to the protein product of the gene of interest are used as probes. A sample is first exposed to an antibody specific for the target protein. A second antibody carrying a fluorescent label is added to allow the location of the first antibody and hence the gene of interest.
- **Chromosome walking** is the final method. In this method, the sequence of long stretches of DNA is determined by overlapping shorter DNA segments. This technique is particularly useful for determining which genes lie next to each other.

Transfection: Transfection refers to the introduction of foreign DNA into an eukaryotic cell. Organisms that have taken up foreign DNA are referred to as **transgenic**. Several methods are used to introduce foreign DNA into eukaryotic host cells. Among the techniques used are **electroportation**, **liposome-mediated transfer**, and **biolistic transfer**. In some cases, defective genes are used to transfect a mouse. If the defective gene replaces the normal gene a **knockout mouse** results.

Restriction Mapping: The size of the fragments produced when a sample DNA is cleaved by restriction enzymes can be used to generate a physical map of the sample DNA. The results of a **partial digest** (one in which the process is stopped before all sites are cleaved) can be compared to those of a complete digest to determine the order of restriction sites. Two enzymes can be used simultaneously to digest a sample of DNA (**double digest**).

Restriction Fragment Length Polymorphisms: Physical changes in the DNA molecule may change the number of restriction sites. As a result, digestion of two samples may yield fragments that differ in size. These differences are called restriction fragment length polymorphisms (**RFLP's**). Differences in restriction sites may also result from differing numbers of repeated short segments termed **variable-number tandem repeats**.

Polymerase Chain Reaction: This technique takes advantage of our knowledge of the replication process to amplify small samples of DNA. Primers complementary to sequences found on either side of the region of interest are created and used to initiate replication. These primers and all materials necessary for replication are mixed with a DNA sample. A series of heating and cooling cycles facilitates DNA denaturation (when the temperature is raised) and replication of the areas between the primers (when the mixture is cooled). The number of DNA fragments increases exponentially with each cycle.

DNA Sequencing: The sequence of bases in a DNA molecule can be determined by manipulating the process of DNA replication. In the **dideoxy method**, bases lacking a free hydroxyl group are used to terminate DNA replication when they are inserted into a growing DNA chain. A sample of DNA is divided into four sub-samples, and each is mixed with a different dideoxy base. As replication proceeds in each sub-sample, fragments of different lengths are produced. Each fragment corresponds to the position of the complementary base in the original DNA molecule. The sequence is revealed in the pattern of fragments produced.

Problem-Solving Hints

Hint 1: Restriction enzymes cleave DNA molecules at the same sequences regardless of the source of the DNA. In terms of the DNA fragments produced, differences between individuals result from differences in restriction site number or placement.

Hint 2: Loss of a restriction site results in the formation of one large fragment in place of two small fragments (which together equal the size of the large fragment). Addition of a restriction site results in the formation of two small fragments in place of one larger fragment.

Hint 3: In restriction mapping, fragments appearing in a double digest should add up to a fragment from a single digest.

Hint 4: Messenger RNA can be converted into double stranded DNA by the enzyme reverse transcriptase.

Hint 5: PCR amplifies the DNA sequence occurring between two primers.

Hint 6: The sequence of a DNA molecule (5′ to 3′) is determined by reading the gel from bottom to top in the dideoxy method.

Practice Problems

Problem 12-1. Examine the autoradiogram of a gel resulting from a DNA sequencing experiment shown below. Determine the base sequence of both strands of the DNA molecule. Provide the banding pattern that would be expected if the complementary strand of DNA were sequenced by the same method.

Target Strand: Complementary Strand:

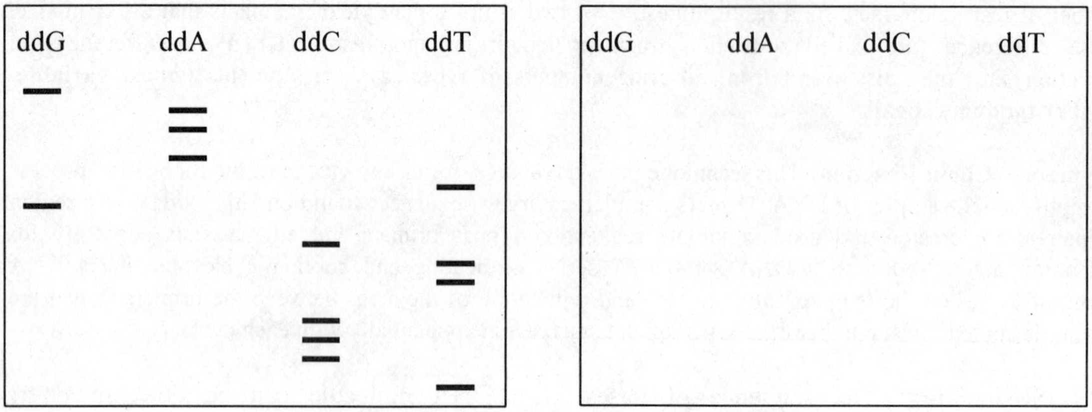

Problem 12-2. A hypothetical gene has three different alleles (A1, A2, and A3). The RFLP sites in the region of the chromosome containing this gene are indicated below. The locations of sequences complementary to the probes used in the Southern blot analysis are also indicated. Predict the pattern of bands that would result from the analysis of homozygous individuals and from the analysis of individuals heterozygous for any pair of alleles.

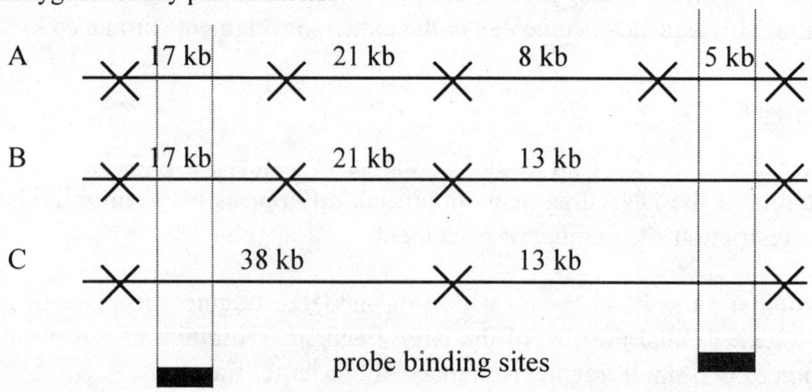

Problem 12-3. Consider a 1250 bp long piece of DNA that is digested with three different restriction enzymes singularly and in all combinations (double digests). The results of each single and double digest are given below. Use the restriction fragment size data to construct a restriction map showing all restriction sites.

Enzyme Treatment	Resulting Restriction Fragment Sizes			
Enzyme A	150	500	600	
Enzyme B	250	1000		
Enzyme C	450	800		
Enzyme A and B	100	150	400	600
Enzyme A and C	150	200	300	600
Enzyme B and C	200	250	800	

Problem 12-4. Given below is the nucleotide sequence or a DNA sample from a hypothetical organism. Also given is the sequence of primers used to amplify a section of this DNA molecule. What sequence will the majority of the PCR products possess?

DNA sequence: 5′ TCCACCCGTCGCAACAAAATTAACGTAACTTTACCCAACA 3′

3′ AGGTGGGCAGCGTTGTTTTAATTGCATTGAAATGGGTTGT 5′

Probes: 3′ TGGG 5′ 5′ ACCC 3′

Problem 12-5. Many of the large genome centers participating in the Human Genome Project can sequence DNA at a cost of \$0.50 per base pair. If all laboratories involved in the project were as efficient, how much would it cost to sequence the entire human genome?

Problem 12-6. Given below is a series of overlapping, contiguous regions of a chromosome. Two cloned genes were found to hybridize to pieces of this contig. Region D is the approximate location of the first gene while the second gene was mapped to region F. To which segments (1-6) will each cloned gene hybridize ?

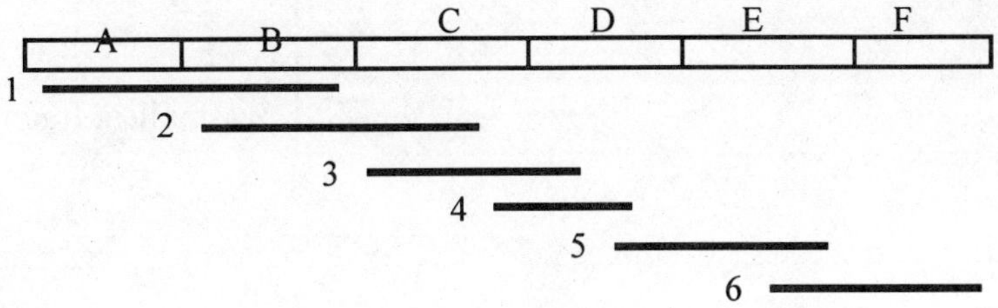

General Questions

Question 12-1. Restriction enzymes are important tools of geneticists. From what natural source are they obtained and of what value are they in nature? Why are they useful in forensic science, for example?

Question 12-2. Discuss the ethical, legal, and social aspects of the human genome project.

Question 12-3. In 1997, cloning organisms from adult cells became a reality with the birth of Dolly. Under what circumstances, and for what purposes, is cloning acceptable in your opinion?

Solutions to Practice Problems

P12-1. Recall that reading the gel from bottom to top reveals the DNA sequence of the 5′ to 3′ strand (*Hint 6*). First, you need to determine the sequence of the target strand (5′ TCCCTTCGTAAAG 3′). Second, determine the sequence of the complementary strand (3′ AGGGAAGCATTTC 5′). Finally, determine the relative lengths of fragments that would result if a dideoxy base were inserted into the growing strand.

Target strand template: 3′ GAAATGCTTCCCT 5′

Fragments ddG sample:

5′ CTTTACG 3′ (7 bp)

5′ CTTTACGAAG 3′ (10 bp)

5′ CTTTACGAAGG 3′ (11 bp)

5′ CTTTACGAAGGG 3′ (12 bp)

Fragments ddC sample:

5′ C 3′ (1 bp)

5′ CTTTAC 3′ (6 bp)

Fragments ddA sample:

5′ CTTTA 3′ (5 bp)

5′ CTTTACGA 3′ (8 bp)

5′ CTTTACGAA 3′ (9 bp)

Fragments ddT sample:

5′ CT 3′ (2 bp)

5′ CTT 3′ (3 bp)

5′ CTTT 3′ (4 bp)

5′ CTTTACGAAGGGA 3′   (13 bp)

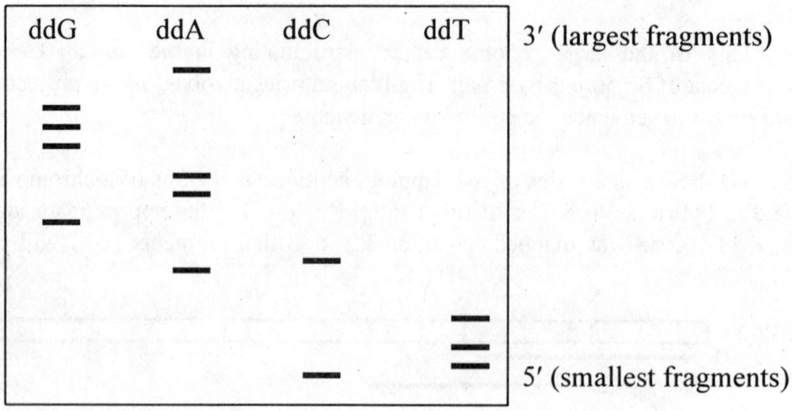

P12-2.

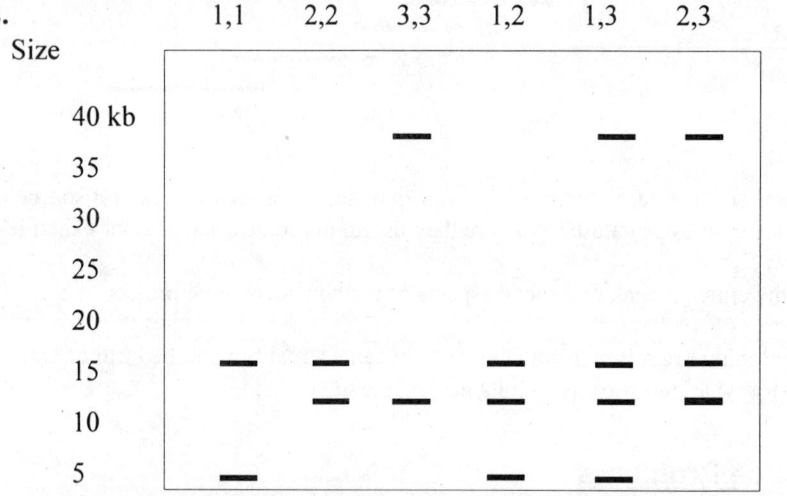

P12-3. To solve this problem, you should begin by determining the number of restriction sites that are recognized by each enzyme. The three fragments produced by enzyme A indicate that there are two restriction sites recognized by this enzyme. Enzymes B and C each produce only two fragments and thus recognize only one restriction site. Now you can compare the results from the single digests to those obtained from double digests to determine the relative distances between restriction sites and to construct the entire restriction map (*Hint 3*). Consider enzymes B and C first. The double digest produces three fragments (200, 250, and 800 bp). Notice that the 250 bp fragment appears in the single digest involving enzyme B and that the 800 bp fragment appears in the enzyme C single digest. This suggests that the 1000 bp fragment (enzyme B) contains the 800 bp fragment (from enzyme C) plus 200 bp from the 450 bp fragment (from enzyme C) (*Hint 2*). The 1000 bp fragment therefore contains a restriction site for enzyme C.

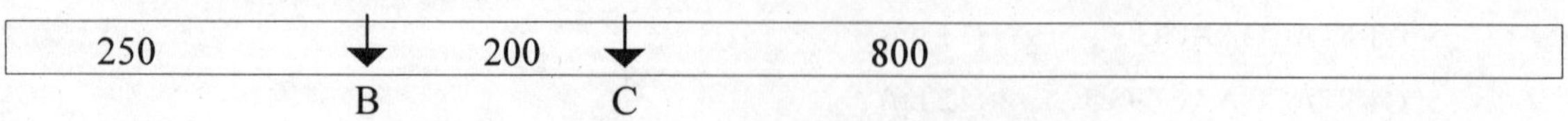

Now consider the results of digestion with enzyme A and B. The double digest reveals four fragments (100, 150, 400, and 600 bp). The 150 and 600 bp fragments are also seen in the single digests. Notice that addition of the 100 and 150 bp fragments (double digest) yields the 250 bp fragment found in digest B. This fragment contains a restriction site for enzyme A. Addition of the 400 and 600 bp fragments from the double digest yields the 1000 bp fragment from the single B digest. The second restriction site is located within this fragment. We can surmise that the 150 bp fragment is on the end, followed by the 100 bp fragment and then the 400 bp and 600 bp fragments. We can make this determination because this arrangement is the only one possible to generate the 500 bp fragment seen in the single digest for enzyme A.

150	↓	100	↓	400	↓	600
	A		B		A	

Finally, the combined map should yield the fragments found in the double digest involving enzymes A and C. The combined map is shown below. As you can see, this map yields fragments consistent with those reported for the double digest indicating we have correctly mapped all restriction sites.

150	↓	100	↓	200	↓	200	↓	600
	A		B		C		A	

P12-4. Remember that the sequence lying between the probes is amplified during the polymerase chain reaction (*Hint 5*). The probe sequences (in the new molecules) are underlined below. The sequence that is amplified is in bold. Very few of the final PCR products will contain the bases occurring before and after the probe binding sites (those in italic).

5′ *TCC*ACCC**GTCGCAACAAAATTAACGTAACTTT**ACCC*AACA* 3′
3′ *AGG*TGGG**CAGCGTTGTTTTAATTGCATTGAAA**TGGG*TTGT* 5′

P12-5. The entire human genome is composed of an estimated three billion base pairs. At fifty cents per base pair, sequencing the entire genome would cost approximately 1.5 billion dollars. Given that approximately 70 million dollars was invested in locating the cystic fibrosis gene alone, the international effort is very cost effective.

P12-6. The clone will hybridize to all fragments overlapping its location. Gene number one will hybridize to fragments 3, 4, and 5 while gene two will hybridize to fragment 6 only.

Answers to General Questions

Q12-1. Restriction enzymes are isolated from bacterial cells where they normally function to cleave foreign DNA. Each restriction enzyme is named for the bacterial species from which it was isolated. Restriction enzymes are extremely useful in recombinant DNA protocols. In particular, the fact that they cleave the DNA at known sites makes them valuable for creating genetic profiles of individuals. Without a way of cleaving the DNA of every individual at exactly the same place, DNA fingerprinting would not be possible (*Hint 1*).

Q12-2 and Q12-3. Questions two and three ask you to consider the social aspects of new genetic technologies. We are living in a time that is sure to be remembered as "the age of genetics." Unfortunately, the technological advances in this field have far outpaced our ability and/or willingness to discuss the social implications and applications of these technologies. Considerable positive outcomes of the human genome project and cloning are possible. For example, these abilities may one day allow us to correct genetic defects prior to conception or birth and to provide new options to infertile couples. However, there are also many potential concerns about this new knowledge and ability. For example, who will have access to genetic information? What will these technologies be used for? Who will make decisions about what the technology can and cannot be used for? Who will the technology benefit? Who will it hurt? The questions are endless. Think about the concerns you have. Also think about the positive outcomes you hope to see in the future. Write each of these down. What factors have shaped your opinions? Talk about these issues with your classmates and your instructor. I have not provided you with a list here because I want you to think about these new technologies and to form your own opinions. These twelve chapters have provided you with the foundation you will need to critically evaluate the progress that has been made and to speculate about the future of genetics.

Answers to Even Numbered End of Chapter Exercises, Problems and Critical Thinking Questions

2. At least three sequences of twofold symmetry (palindromes) exist (with shorter embedded sequences): GAATTC, GGATCC, GCATGC.

4. A particular restriction endonuclease is unsuited for cloning if it destroys the DNA of interest or if it does not have an appropriate site in the vector or surrounding the foreign DNA.

6. The heteroduplex would look similar to the one in figure 12.19.

8. The steps in creating cDNA from messenger RNA are shown in figure 12.10. Radioactive cDNA and messenger RNA can be obtained either by using radioactive triphosphate nucleotides during the synthesis of the cDNA and messenger RNA or by end-labeling the products.

10. Use radioactive alanine transfer RNA as a probe in either a whole digest (Southern blotting) or in a genomic library (dot blotting).

12. If a messenger RNA could not be isolated, a probe could be constructed using the codon dictionary if the amino acid sequences of the protein product were known. A genomic library could be constructed with the intent of looking for the location of the gene by its expression. Expression vectors are those whose foreign DNA is expressed.

14. Foreign DNA can be introduced into eukaryotic cells by infection with animal virus vectors as well as several other techniques including electroporation, liposomes, and biolistics. Transfection is the introduction of foreign DNA into eukaryotic cells. A transgenic mouse has been successfully transfected; it has incorporated the foreign DNA into its genome.

16. The cells picked up two plasmids, one a single ligated plasmid (colony 1) and one that is a dimer (or higher *mer*) of the plasmid (colony 2). The second band moves more slowly than the original band; it must be larger than the original. Two or more molecules with the same sticky ends can combine to form one larger molecule.

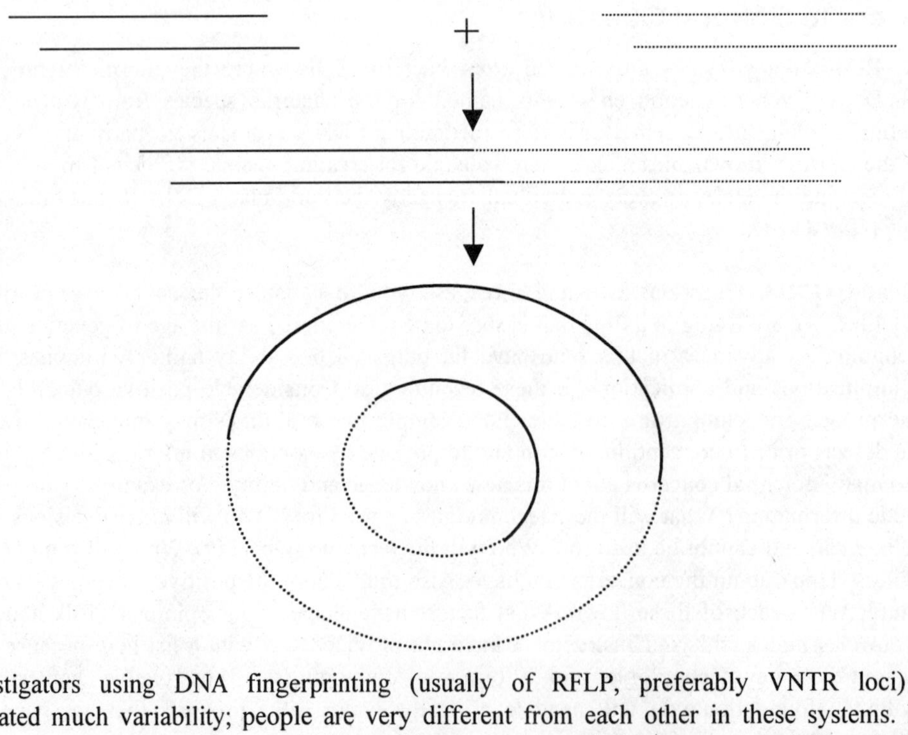

18. Investigators using DNA fingerprinting (usually of RFLP, preferably VNTR loci) have demonstrated much variability; people are very different from each other in these systems. Hence people can be identified with very high confidence. These techniques are especially suitable for paternity exclusion. RFLPs of mother, child, and putative father are examined. The father's genome must include (barring mutation) all the bands in the offspring not contributed by the mother.

20. The original was cut between the following lengths: 50-200-100-400.

22. Alternative II is correct. Note that among the clues, alternative I requires a band of a segment that is 200 bp long.

24.

3 ↓ 5 ↓ 4.2 or

3 ↓ 4.2 ↓ 5 or

5 ↓ 3 ↓ 4.2

A linear molecule with *n* sites will produce *n* + 1 fragments, so there are two sites within the molecule. Any one of the three fragments could be in the middle.

26. The molecule either has no *Eco*RI sites, or it is a circle. A circle with *n* sites yields *n* molecules. If there are no *Eco*RI sites, and the molecule is linear, we expect to see the same number of *Bam*HI fragments from both the single *Bam*HI digest and from the double digest. Since we got four fragments in the double digest, but only three in the *Bam*HI digest, the molecule must be a circle. The 2.0 kb and 4.5 kb fragments appear for both *Bam*HI and *Bam*HI + *Eco*RI, so they must not contain an *Eco*RI site. The 5.0 kb fragment must contain an *Eco*RI site 2.5 kb from one end.

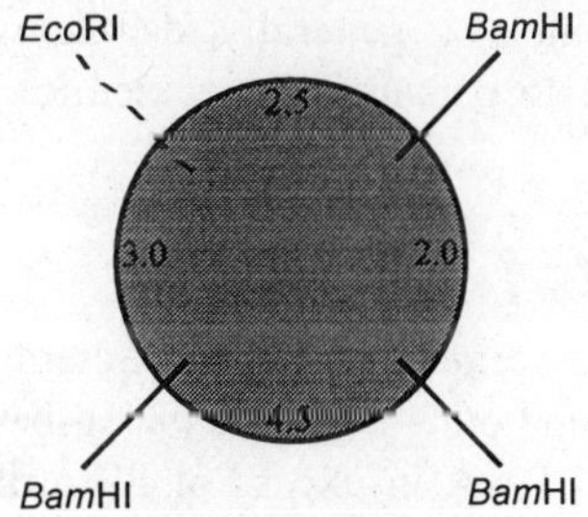

28.

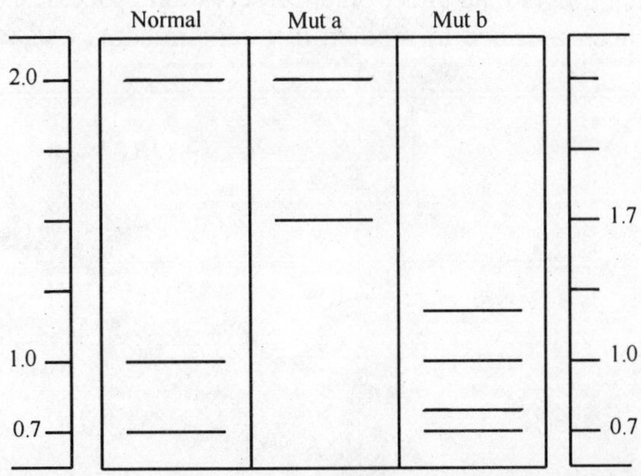

The first mutant will be cleaved only once, yielding bands of 1.7 kb and 2.0 kb. The second mutant will produce the 1.0 kb and the 0.7 kb fragments since this site is conserved. A new site within the 2.0 kb fragment will generate two new fragments, the sum of which is 2.0 kb. Thus the loss of a site produces two bands, and the acquisition of a new site yields four bands.

30. 0.5, 1.0, 2.0, 3.5 kb. The 1.0 kb band should have a higher intensity than the other bands. It helps to redraw the *Eco*RI sites:

*Eco*RI	0.5	↓	1.0	↓	3.0	↓	3.5
*Hind*III			2.5	↑	5.5		

Since *Hind*III does not cut the 3.5 band, it must cut within the 3.0 band, as drawn here. There will be two 1.0 kb bands produced as well as bands of 0.5, 2.0, and 3.5 kb.

32. Polymerase chain reaction (PCR) is a rapid method to amplify DNA segments. It is particularly useful when only a small amount of the DNA is available and primers flanking the region can be created.

34. 5′-CAATAGGTCGAGGTTCAATGG-3′

36. Overlap can work for one, four, or seven bases. Lowercase letters indicate the stop sequence (tga), boldface indicates the **ATG** start sequence, and X indicates any base. One-base overlap, … tga**TG** … ; four-base overlap, … **At**ga … ; seven-base overlap, … **ATG**Xtga.…

38. 2-*mer,* 5-*mer,* 11-*mer,* 12-*mer,* 13-*mer.* Recall that the dideoxy method produces bands that result from the synthesis of a new complementary band. This strand will have the following sequence:

5′___CTGATAAGGCTTTG___3′
↑ ↑ ↑↑↑

Dideoxythymidine will produce fragments wherever a T appears in the new strand (arrows).

40. Benefits of gene cloning are described for medicine, agriculture, and industry. Many people are concerned that toxic, carcinogenic, or generally damaging genes will be introduced into the environment or pass from person to person in an epidemic. The Human Genome Project provokes fears of invasion of privacy.

Critical Thinking Question

2. For several years the premise seemed reasonable. Scientists began to extract DNA from fossils, including the DNA from termites and other organisms in amber. The DNA was then amplified with the polymerase chain reaction. Fossil DNA in excess of 30 million years old was reported. However, recently, scientists have cast doubt on these claims. DNA begins to break down immediately on the death of the organism from various processes; the possibility of DNA millions of years old and still intact is now considered highly unlikely, although one group claims that DNA preserved in amber retains its integrity much longer than under other preservation processes. The DNA results reported for the fossil DNA have been ascribed to modern-day contaminants and to artifacts of the polymerase chain reaction.

CHAPTER 13
GENE EXPRESSION: CONTROL IN PROKARYOTES AND PHAGES

Key Genetic Concepts

Control of Transcription: While some proteins are produced at all times, most proteins are produced only as required by specific conditions. The **operon**, a sequence of genes controlled by a single **operator**, is a unit of coordinated gene expression. Proteins needed only in certain environments are generally controlled by **inducible operons**. Transcription of the genes encoding these proteins are "turned on" when they are needed. Transcription of genes encoding proteins that are no longer needed can be repressed. In this case, gene expression is controlled by **repressible operons**. The major types of transcriptional control are outlined below.

Inducible Systems: The **lac operon** is an example of an inducible system. Production of the enzyme beta-galactosidase, which splits lactose into glucose and galactose, is produced only when lactose is the available energy source. The availability of nutrients (i.e. glucose and/or lactose) controls the transcription of the genes (**structural genes**) for β-galactosidase production and two other necessary enzymes. In the presence of lactose, the gene is "turned on" and transcription of the enzyme proceeds (RNA polymerase can freely bind to the DNA). When lactose is not present, the enzyme is not needed and the gene is "turned off." Though not transcribed, the promoter is required to "position" the RNA polymerase for downstream transcription of the operator gene and the structural gene for β-galactosidase. In this case, a **regulatory gene** produces a protein (the **repressor**) that binds to the operator gene thereby blocking the attachment of RNA polymerase. If RNA polymerase cannot bind, transcription is inhibited. When lactose is present, it binds to the repressor protein causing it to change its shape. The repressor then releases from the operator gene and the blockage is removed. Mutations can alter the function of the regulatory gene or the operator gene. Operator mutations are **cis-dominant**: they affect only the operon in which they are located. Regulator mutations are **trans-acting**: they are recessive to a wild-type regulator gene regardless of which operon is involved.

Repressible Systems: In a repressible system, synthesis of an enzyme or group of enzymes is turned off when they are no longer needed. The **tryptophan operon** is an example of a repressible operon. The structural gene in this operon codes for an enzyme essential to the biosynthesis of the amino acid tryptophan. In repressible systems, the inactive repressor is normally made. This protein cannot recognize the operator unless it is in combination with tryptophan (a **corepressor**). An excess of tryptophan "turns off" transcription by forming a **corepressor-repressor complex** which binds to the operator gene thereby blocking the attachment of RNA polymerase.

Attenuation Systems: The tryptophan operon is also regulated by a mechanism called attenuation. The attenuator is a nucleotide sequence located between the promoter and structural gene of the operon. This sequence acts to regulate the movement of RNA polymerase. The rate of *translation* determines whether or not the tryptophan genes are transcribed. With high levels of tryptophan, translation speed increases and transcription is terminated. When levels of tryptophan are low, translation slows and transcription is completed.

Lytic and Lysogenic Cycles in Phage λ: Control of gene expression is also important in bacteriophages. The bacteriophage λ can enter either a **lysogenic** or **lytic** cycle. Which cycle is entered depends upon which repressor accesses the operator site. Lysogeny is favored by the CI repressor while the lytic cycle is favored by the Cro repressor. A complete list of the gene products involved in phage λ infection including promoters, terminators, and antiterminators is provided in Table 13.1 of your textbook.

Transposable Genetic Elements: Transposons are regions of the DNA that can move from one place to another in the genome. Alternatively, a transposon can insert a copy at a new place with the original still existing at the original site. Insertion sequences, or IS elements, were the first transposons identified in bacteria. Composite transposons consist of a central region surrounded by two IS elements. The exact mechanism of transposition is, as yet, unknown. Transposition can have a variety of effects on the phenotype and genotype. Insertion of a transposon can disrupt gene expression, prevent gene expression, or cause deletions or inversions.

Translational Control of Gene Expression: Translational control mechanisms may influence the efficiency of mRNA translation. RNA-RNA hybridization can also function as a regulatory mechanism. If both strands of the DNA molecule are transcribed, the resulting mRNA's will possess complementary sequences. Regulation could be achieved if the two complementary mRNA's hybridize forming a double-stranded molecule that could not be translated. The regulating mRNA is referred to as an **antisense** RNA. The nucleotide sequence found in the mRNA molecule can also influence the rate of translation. **Codon preference** results if the mRNA of a gene contains codons specifying more common transfer RNA molecules. A greater number of preferred codons results in increased translation while the process is slowed by the presence of codons specifying rarer transfer RNA's. A final translational control mechanism is the **stringent response**. When a bacterial cell is starved of amino acids, a protein (**stringent factor**) is produced. This factor causes transcription of mRNA, rRNA, and tRNA molecules to cease.

Posttranslational Control: Control can be exerted posttranscriptionally by feedback inhibition. In this case, the product of the biosynthetic pathway turns off the pathway. Posttranslational control can also be achieved through the differential degradation of mRNA's. Those mRNA's coding for gene products needed in large quantities have a longer life than those coding for products needed less often.

Problem-Solving Hints

Hint 1: A mutated repressor element can disrupt expression of any gene in the genome (trans-dominant) while a mutation in a promoter or operator will affect only genes on the same chromosome (cis-dominant).

Hint 2: A co-repressor must bind with an inactive repressor to control a repressible operon.

Hint 3: Mutations of structural genes result in the production of defective products (enzymes or other proteins).

Hint 4: Attenuators are sensitive to the speed of translation. This type of control is possible because transcription and translation are coupled in prokaryotic systems.

Hint 5: When determining operon functioning, assign phenotypes to the nuclear DNA and plasmid DNA separately before considering their joint action.

Practice Problems

Problem 13-1. Of what consequence are mutations in the operator, promoter, and repressor genes of the lac operon?

Problem 13-2. If a cell is infected with λ phage possessing a defective C1 genes, will lytic or lysogenic development occur? Which will occur if the phage possesses a defective Cro gene?

Problem 13-3. Consider a sequence of genes, A B C D, in a hypothetical inducible operon. Given the consequences of mutation (below), identify which gene is the promoter, the operator, the structural gene encoding enzyme 1, and the structural gene encoding enzyme 2.

Mutation	Enzyme 1	Enzyme 2
A	inducible	none
B	constitutive	constitutive
C	none	inducible
D	none	none

Problem 13-4. On the DNA molecule provided below, indicate the location of the RNA polymerase, the repressor protein, and the mRNA molecule during normal transcription of the lac operon and during inhibition of transcription.

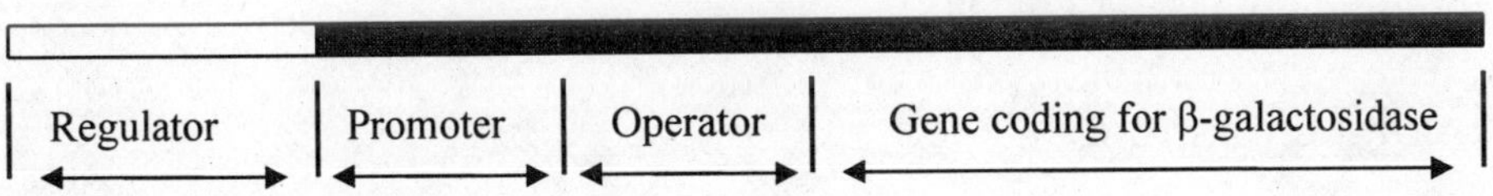

General Questions

Question 13-1. Why is there a need to regulate gene expression?

Question 13-2. Is the term "jumping gene" an accurate description of all bacterial transposable elements?

Question 13-3. Attenuation systems are found in prokaryotic organisms but not in eukaryotic organisms. Why?

Solutions to Practice Problems

P13-1 A mutation in the operator region of the operon prevents binding of the repressor protein (*Hint 1*). As a result, transcription is constitutive and the operon is always "on." A mutated promoter will inhibit positioning of the RNA polymerase. As a result, transcription is prevented regardless of the presence of a wild-type promoter on the plasmid. A mutated repressor cannot bind to the operator resulting in constitutive production. A wild-type repressor gene on the plasmid will restore the inducible function of the operon provided transcription is constitutive.

P13-2. The C1 gene is associated with a lysogenic cycle while the Cro gene is associated with a lytic cycle. If the C1 gene is defective (a), lytic development will occur. Lysogenic development will occur in (b).

P13-3. Genes A and C are the structural genes (*Hint 3*): gene A codes for enzyme 2 and gene C codes for enzyme 1. Gene D is probably the promoter, and Gene B is probably the operator.

P13-4.

Normal transcription of enzyme β-galactosidase:

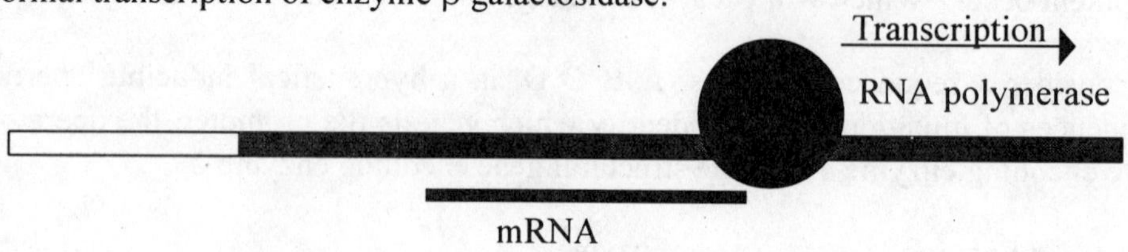

Inhibition of β-galactosidase transcription:

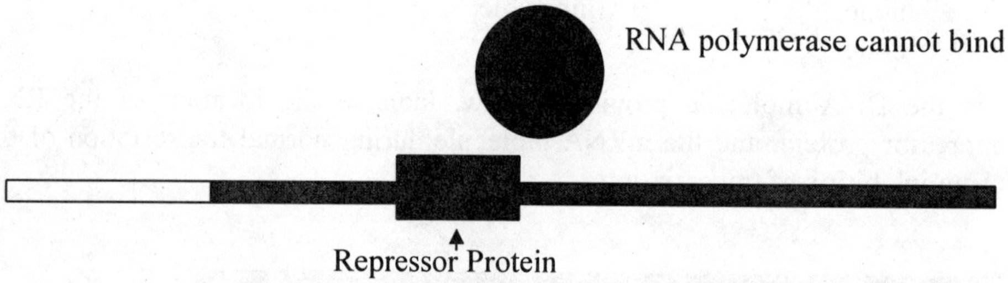

Answers to General Questions

Q13-1. Control of expression is important for several reasons. First, without control processes, the cell would produce all of its proteins in large quantities whether they were needed or not. The cell requires approximately 1200 high-energy bonds per protein. Therefore, this energy-consuming process would be costly to the cell. Second, there may be times when certain proteins are needed in greater or lesser quantity: there must be a way of suppressing or enhancing the activity of specific genes. Finally, control of gene expression in eukaryotic organisms is an important aspect of cell differentiation (see the next chapter).

Q13-2. The description "jumping gene" is not accurate in all cases. While some transposable elements excise from the DNA and insert in a new location, other elements insert copies in other areas. This second type of transposable element remains at the original site and therefore does not "jump."

Q13-3. In attenuation systems, the speed of translation controls transcription (*Hint 4*). This requires the processes of transcription and translation to be coupled, as they are in prokaryotic cells. In eukaryotic cells, transcription takes place in the cell nucleus while translation occurs in association with the ribosomes in the cytoplasm.

Answers to Even Numbered End of Chapter Exercises, Problems and Critical Thinking Questions

2. a.–d. inducible; e. constitutive

4. The most likely candidates are: (1) mutations in the repressor such that the repressor can never bind to the operator (i^-), and (2) mutations in the operator such that repressor is unable to bind (o^c). The partial diploid for the first class can be indicated as $i^-o^+p^+z^+y^+/i^+o^+p^+z^+y^+$. Since the second DNA provides a good repressor, which is *trans-acting,* both DNAs will be regulated, and inducible by lactose. The second mutant, $i^+o^cp^+z^+y^+/i^+o^+p^+z^+y^+$, contains an operator that cannot bind the repressor. This o^c mutation is *cis-dominant,* and the left DNA will always be on.

6. The operator is probably between S and T, and the promoter is probably to the left of Q. Deletions of the operator sequence will prevent binding by the repressor, and the operon will always be on. This result is seen in 3 and 4. Both strains 3 and 4 are missing part of the region between S and T. A promoter deletion will never make RNA, and this is seen in deletion 1.

8. An operon with a CAP site (e.g., arabinose in *E. coli*) has enhanced transcription when glucose is absent (positive control). However, these operons usually are inducible also. Lack of an induction (negative control) mechanism is very inefficient.

10. $p\ o^c\ e_1 \ldots a^+/\text{F}'p\ o^+\ e_2 \ldots a^+$. Operator constitutive mutants (o^c) are *cis* dominant; the e_1 operon cannot be repressed. The e_2 operon is repressible. Regulator mutants (i^-) are recessive in that a wild-type regulator (i^+) can control all operons in a merozygote. This type of control can work for any synthetic pathway in which an end product or derivative can be a corepressor.

12. a. No transcription; *cis*-dominant. b. Yes, transcription; recessive to wild-type. c. Yes, transcription; recessive to wild-type. If polymerase cannot bind, no transcription or translation can occur, but this effect will only be seen in the one DNA strand; wild-type DNA will be able to bind RNA polymerase. If the repressor-corepressor complex cannot bind to DNA, RNA polymerase will be able to bind, and the operon will always be on. A good repressor, regardless of source, will be able to bind both DNAs. If the repressor cannot bind transfer RNA, it cannot bind to the operator, and the operon will always be on. In partial diploids, the wild-type repressor, bound to the transfer RNA, can repress both operons.

14. Repression; lytic response

16. Assuming that the operator mutants prevent the operator from recognizing repressor: o_{R1}, should result in the lytic response; o_{R3}, should result in lysogeny. Assuming that promoter mutants do not recognize RNA polymerase: p_L, no action by phages possible; p_{RE}, lytic response; p_{RM}, lytic response possible; p_R, no response by phage because no *cII* and no right or late operon production; t_{L1}, read-through transcription of left operon even if *N* product absent, lysogeny or lytic response possible. t_{R1}, read-through transcription of right and late operons even without *N*-gene product; should result in lytic or lysogenic response. Mutation of *nutL* results in termination of transcription in left operon resulting in lytic response. Mutation of *nutR* results in termination of transcription in right operon resulting in possibility for integration but not lytic cycle.

18. UV damage induces DNA repair systems. The RecA protein, involved in repair, also cuts the λ repressor protein, leading to induction of the lytic cycle.

20. Lysis results. Since *cI* controls transcription of O_L and O_R, it will not be able to bind to O_L and O_R, and transcription will occur, leading to the lytic cycle.

22. IS elements have been called selfish DNA because they replicate copies of themselves in the host genome without necessarily providing the host with any obvious benefit.

24. See figure 13.32. Transposase is involved in joining nonhomologous single-stranded DNA. Resolvase catalyzes a recombinational event that separates the two elements of a cointegrate. DNA polymerase I and ligase function in processes that repair small patches of single-stranded DNA during transposition.

26. See figure 13.33.

28. A polar mutation is one, usually within an operon, that prevents the expression of a distal gene. It can be caused by a deletion or the formation of a transcription stop signal through mutation or from within a transposon.

30. Transcriptional control is much more efficient; it occurs before a lot of energy is wasted on transcription or translation.

32.

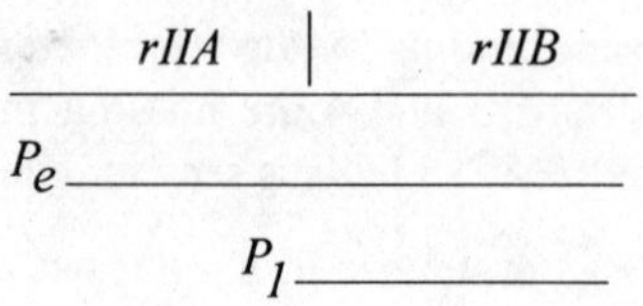

The fact that early products are present in equimolar amounts suggests that early mRNA is polycistronic. This suggestion is confirmed by the nonsense mutations in *rIIA* that behave as polar mutations in an operon. One way to explain the excess *rIIB* in late infections is to assume that *rIIB* has a separate promoter for late transcription. The *rIIA* deletions support this notion and suggest that the *rIIB* promoter is in fact located within *rIIA*.

34. Antisense RNA is complementary to messenger RNA. It can have regulatory functions by forming a double helix with (and thus inactivating) its functional complement. An obvious source of antisense RNA is transcription of the complementary DNA strand of the strand normally transcribed. Another source could be from the transcription of a gene that has evolved from the original gene and thus has a similar base sequence. Clinically, it might prevent functioning of RNAs that have disease functions.

36. Protein degradation can be controlled by the nature of the N-terminal end or by PEST regions (regions rich in proline, glutamic acid, serine, and threonine). (See table 13.3.)

Critical Thinking Question

2. It seems inefficient for the bacterium to have such short-lived messenger RNAs and proteins. However, that is the price the cell pays for the ability to adapt very quickly to a changing environment. If messenger RNAs and proteins were long lasting in the cell, the cell's metabolism would reflect decisions (transcription and translation) made a longer distance in the past, not the current circumstance. Actually, the fact that the messenger RNAs and many proteins do turn over quickly predicts that the bacterium does find itself often in a very rapidly changing environment. That is consistent with both free living in an aqueous environment or living in the intestines of a higher organism.

CHAPTER 14
THE EUKARYOTIC CHROMOSOME

Key Genetic Concepts

Eukaryotic Chromosomes: Each eukaryotic chromosome contains one double helix of DNA (**uninemic**). The DNA and histone proteins are integrally associated and arranged in regularly spaced spherical structures called **nucleosomes**. Between nucleosomes are stretches of **linker DNA**. This entire structure is further coiled. The **scaffold** structure of the chromosome is formed by nonhistone proteins. **Polytene chromosomes** result from replication and synthesis of homologous chromosomes without cell division. The dark staining regions of chromatin condensation that occur on eukaryotic chromosomes during early mitosis or meiosis are called **chromomeres**. In polytene chromosomes, several chromomeres aligned laterally will appear as a dark band. The diffuse, uncoiled regions in polytene chromosomes represent regions where transcription is actively occurring and are referred to as **chromosome puffs**. **Lampbrush chromosomes** have a looped-out configuration due to the elaboration of mRNA and are found in amphibian oocytes. The DNA that is transcribed is referred to as **euchromatin**. Chromatin that does not contain active genes is called **heterochromatin** while the repetitive DNA that surrounds the centromere is referred to as **constitutive heterochromatin**.

Telomeres: The regions at the ends of a chromosome are referred to as **telomeres**. Most are repetitions of a 5-8 base sequence and are required for chromosome stability and replication. These sequences are added to chromosomes by the enzyme **telomerase**. In the cells of most higher organisms, the telomeres shorten with each cell division because telomerase is not active. Telomeres are believed to be the "time clock" controlling cell division.

Repetitive DNA: An estimated 97% of eukaryotic DNA does not code for RNA (junk DNA). Repetitive DNA contains many copies of the same nucleotide sequence. Some repetitive DNA contains multiple copies of a gene. Approximately 10% of human DNA contains highly repetitive nucleotide sequences, 15% contain intermediately repetitive sequences, and 75% contain unique sequences.

Problem-Solving Hints

Hint 1: The midpoint of a plot of cot values versus quantity of remaining single-stranded DNA estimates the length of unique DNA in the sample.

Hint 2: Eukaryotic chromosomes are linear while bacterial chromosomes are circular.

Hint 3: The chemical composition of specific chromosomal regions account for differences in chromosome staining.

Practice Problems

Problem 14-1. Which of the features below are associated with bacterial chromosomes? Which are found in eukaryotic chromosomes? Which of the features are characteristic of both cell types?

a. histone proteins

b. pentose sugars

c. nitrogenous bases

d. telomeres

e. supercoiling

f. non-histone protein scaffolds

g. linear chromosomes

h. circular chromosomes

i. nucleosomes

Problem 14-2. Consider a hypothetical organism with a haploid genome consisting of 10^8 nucleotide pairs. Eighty five percent of the genome is unique-sequence DNA and each gene is represented once. If the average gene sequence is 10^2 nucleotide pairs, how many different genes are potentially encoded in the region described above?

Problem 14-3. One third of the length of a metacentric eukaryotic chromosome consists of satellite DNA. The G-band staining pattern is shown below. The centromere is represented by an open circle. Show the expected R-band and C-band patterns for this chromosome.

Problem 14-4. Shown below are a pair of homologous chromosomes from a karyotype stained with the Giemsa banding procedure. The centromere is represented by an open circle. Would the banding patterns observed for this pair of chromosomes be useful in identifying potential causes of mental retardation in the individual from which they came ?

Maternally derived chromosome

Paternally derived chromosome

General Questions

Question 14-1. Which of the following types of sequences are most common in human nuclear DNA? (a) unique sequences functioning as genes; (b) introns; (c) repetitive sequences.

Question 14-2. In many yeasts, protozoa, and other single-celled organisms, senescence is unknown. This is not true of cells from higher organisms. Suggest a possible explanation for this difference.

Question 14-3. What is the function of the centromere?

Solutions to Practice Problems

P14-1. Pentose sugars (b), nitrogenous bases (c), and supercoiloing (e) are characteristic of both prokaryotic and eukaryotic chromosomes. Only the prokaryotic chromosome is circular (h). Histone proteins (a), telomeres (d), scaffolds (f), linear structure (g), and nucleosomes (e) are all associated with eukaryotic chromosomes.

P14-2. Size of unique-sequence area = 0.85×10^8 base pairs

genes = 0.85×10^8 base pairs / 1×10^2 base pairs per sequence

= 850,000 or 8.5×10^5 different sequences.

P14-3.(*Hint 3*)

R-band pattern 

C-band pattern 

P14-4. Notice that the banding pattern of the maternally and paternally derived chromosomes are not identical: the paternal chromosome is missing the second band on the short arm. The missing band could represent a deletion on the paternal chromosome. Alternatively, the paternally derived chromosome could represent the normal situation with the extra band on the maternally derived chromosome resulting from a duplication. This information would be very useful in diagnosis of a mental deficiency or any other clinical symptom. The G-banding staining technique is the standard technique used by laboratories for cytogenetic study.

Maternally derived chromosome

Paternally derived chromosome

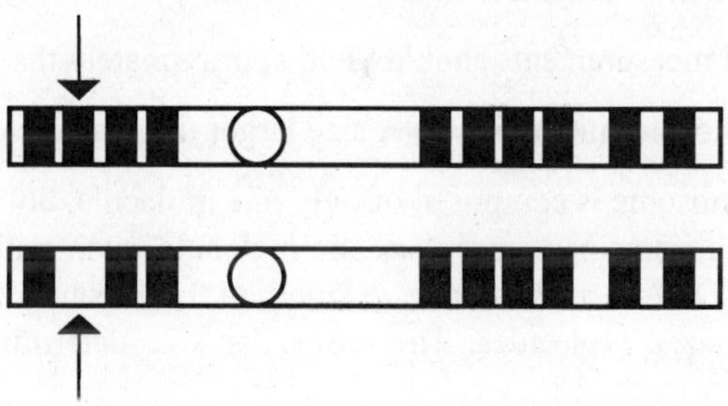

Answers to General Questions

Q14-1. Most human nuclear DNA is made up of repetitive sequences. The function of these repeating sequences is unknown.

Q14-2. Telomerase is active in yeasts, protozoa, and other single-celled life but is not active in the cells of higher organisms. The loss of telomeres is thought to function as a "time clock" of cellular division in higher organisms. Some of the telomere is lost during each division of the cell. It has been suggested that senescence results when a significant proportion of the telomere has been lost.

Q14-3. The centromere serves as the attachment point for the spindle fibers during nuclear division.

Answers to Even Numbered End of Chapter Exercises, Problems and Critical Thinking Questions

2. The evidence for the uninemic nature of the eukaryotic chromosome is summarized in figures 14.1, 14.2, and 14.3.

4. The major protein components of eukaryotic chromosomes are the histones, which form nucleosomes, compacting the DNA, and nonhistone proteins, which form the scaffold and have other functions.

6. The nucleosome is composed of two each of histones H2A, H2B, H3, and H4. Histone H1 seems to be associated with DNA as it enters and exits the nucleosome.

8. G-bands and C-bands are illustrated in figures 14.18 and 14.19. R-bands are the reverse of G-bands. Band structures are summarized in table 14.3.

10. See figure 14.13.

12. A chromosomal puff can be stage-specific, tissue-specific, constitutive, or environmentally induced. It is an area of active transcription in a polytene chromosome.

14. A kinetochore is a protein-RNA structure that connects the chromosome to the microtubules of the spindle. It occurs in a visible constriction of the chromosome called a centromere. Yeast centromeres are about 220 base pairs long with three consensus regions (see fig. 14.22).

16. Cot is a renaturation scale of DNA (concentration × time); see figure 14.30 for various cot curves. $Cot_{1/2}$ values have been used to construct a scale of unique DNA length in DNA samples. The repetitiveness of mouse DNA can be determined from the cot curve of figure 14.31, which indicates three different fractions.

18. Construct a radioactive Alu probe. Then use this probe to locate Alu sequences on the chromosomes by hybridization and autoradiography or fluorescent techniques (e.g., fig. 14.19).

20. See figures 14.35 and 14.36.

22. a. Both measurements should yield approximately the same size molecule.

b. Method 1 will yield a DNA size larger than that determined by method 2.

If the chromosome is composed of only one molecule, both methods give total DNA per chromosome. If the chromosome has more than one molecule, each molecule (method 2) will be less than the total amount of DNA per chromosome (method 1). Assume for example that the chromosome has two identically sized molecules. The molecular size determined by method 2 will be one-half the size calculated by method 1.

24. Nucleases will not digest DNA to which proteins, in this case histones, are bound. Approximately two hundred base pairs must be wrapped around proteins, and there must be some unprotected regions between these two hundred base pair regions. Multiples of two hundred base pairs appear because the nuclease does not cut at every unprotected sequence.

26. Yes, but only the smallest genes. The average coding part of a gene has about one thousand base pairs. A two hundred base pair region would give maximally a protein of sixty-six amino acids. Most proteins have much more than sixty-six amino acids.

28. Repetitive regions must be separated by nonrepetitive regions. If the repetitive sequences are about four hundred base pairs long, the four hundred base pair renaturation mix will contain

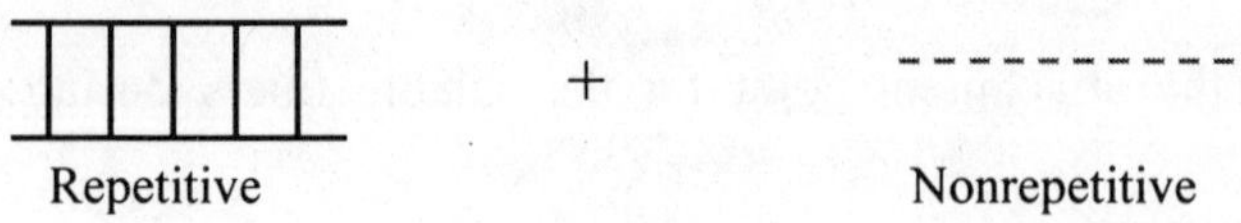

in about equal amounts. The four thousand base pair mix will contain mostly molecules that look like this:

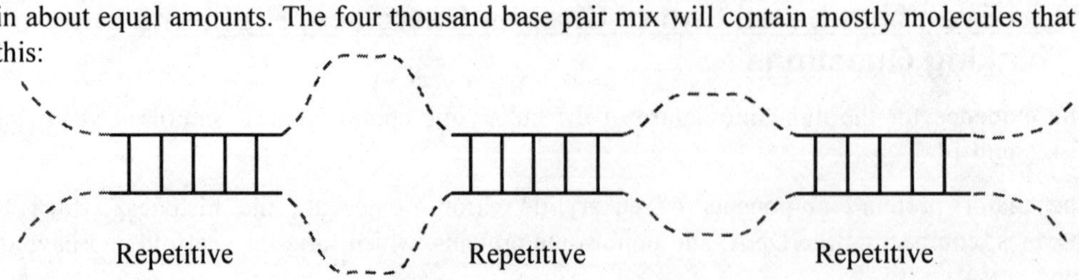

In separating single-stranded and double-stranded molecules, this type of molecule will behave like a double-stranded DNA. We get an unusually high amount of "apparent" renaturation.

30.

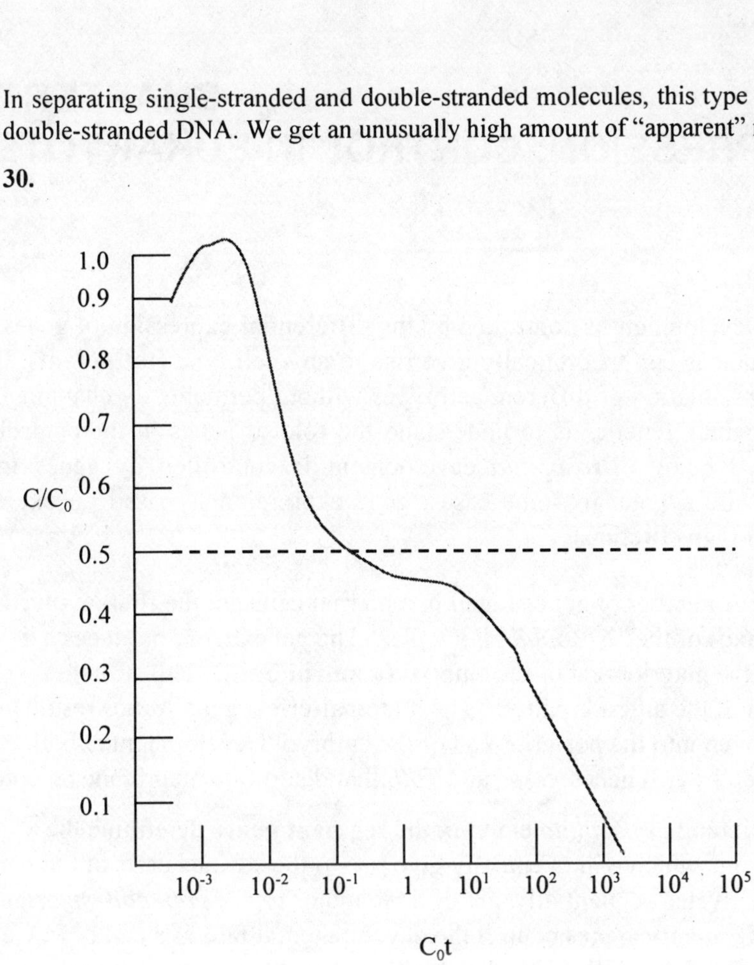

32.

cot	c/c0
0.01	1/1.01 = 0.99
0.05	1/1.05 = 0.95
0.1	1/1.1 = 0.91
0.5	1/1.5 = 0.67
1.0	1/2 = 0.5
5.0	1/6 = 0.17
10	1/11 = 0.09
50	1/51 = 0.02
100	1/101 = 0.01

Critical Thinking Question

2. In most cell types in higher eukaryotes, the number of divisions is finite. In these cells, telomerase is inactive. As the divisions proceed, the telomeres shorten until a length is reached that somehow signals the cell to stop dividing. If a mutation arose such that telomerase became active in an inappropriate cell type, the telomeres could have a functional length reestablished each generation, signaling the cell to keep dividing. The result of uncontrolled cell growth is cancer.

CHAPTER 15
GENE EXPRESSION: CONTROL IN EUKARYOTES

Key Genetic Concepts

Patterns of Development: Development is controlled by the differential expression of genes. In eukaryotic organisms, any nucleus can theoretically give rise to any cell type (**totipotent**). Cell differentiation results in the production of different cell types without permanently changing the DNA. The goal of developmental genetics is to understand the role of genes in the control of development. As summarized below, *Drosophila* development is controlled by genes from maternal cells and genes of the zygote. In some cases, gene expression is based on external environmental factors (**signal transduction**).

Maternal-Effect Genes: A number of genes found in maternal cells are the first involved in determining the general axes of the *Drosophila* body plan. The gene *bicoid* produces a protein that diffuses into the anterior end of the embryo (a **morphogen**). This protein is involved in the formation of the anterior pattern. The maternal-effect gene *Nanos* results in the diffusion of a morphogen into the posterior end of the embryo. Development at both ends of the embryo is controlled by two genes, *torso* and *Toll*, that determine membrane receptors.

Zygote Genes: In a developing *Drosophila* embryo, the **segment genes** determine the number and fate of segments which will eventually give rise to the various parts of the body. These genes, which are activated sequentially, are of three major types: *gap*, *pair-rule*, and *segment-polarity genes*. **Homeotic genes** control the developmental fate of a cell type. Cells with a mutation in a homeotic gene follow the developmental pathway of another cell type. Genes within the *bithorax* complex control the fate of posterior thorax and abdomen development while genes within the *antennapedia* complex control the fate of anterior development. The **homeo box** is a consensus sequence found in *Drosophila* homeotic genes and other developmentally important genes including those of human beings.

Plant Development: Mechanisms of development in plants are similar to those in animals. A number of genes, including floral-meristem identity genes and homeotic genes, are involved in the sequence of steps in floral and pattern formation.

Control of Transcription: Methylation plays a role in controlling transcription in eukaryotes. Inactive genes are generally more fully methylated than active genes. **Transposons** can also affect transcription in eukaryotic cells. Barbara McClintock received a Noble Prize for her pioneering research on transposons in corn. McClintock examined a mutation in maize which causes a kernel to be purple instead of the wild-type color, white. However, she found that not all kernels with the mutation expressed it: the gene causing purple color was sometimes suppressed. McClintock hypothesized that a second gene, which she termed the dissociator or Ds, was responsible for the suppression. The revolutionary idea that McClintock proposed was that the Ds gene moved around the genome, "turning on" or "turning off" adjacent genes. The movement of the Ds gene was controlled by a third gene, the activator or Ac gene.

Immunogenetics: Antibodies or immunoglobulins (Igs) protect an organisms from substances that evoke an immune response (**antigens**). Antibodies, which are produced by B lymphocytes, are composed of two light and two heavy polypeptide chains. The constant region of the chains recognize specific antigens. Mutations of this region are called **allotypes** and are inherited in a

Mendelian fashion. In the variable region, each individual may possess more than two alleles (**idiotype variation**). This variability results in a great diversity of antibodies. Similar diversity is observed in T-cell receptors which recognize major histocompatibility complex (MHC) proteins.

Cancer: Cancer is now recognized as a genetic disease that results from the inappropriate activity of certain genes. Most cancers arise from a single aberrant cell that proliferates (cancer is **clonal**). This inappropriate activity results when genes are mutated or when genes are imported or activated by viruses. A predisposition for cancer formation may be inherited.

Tumor-Suppressor Genes: Tumor-suppressor genes normally act to suppress malignant growth. However, if mutated, the homozygous recessive genotype results in a loss of gene function and the formation of cancer. Examples of tumor-suppressor genes include the genes for retinoblastoma, Wilm's tumor, and **p53**. A normally functioning p53 gene will trigger programmed cell death (**apoptosis**) if damage to the DNA is severe. Approximately 50% of cancers involve loss of p53 activity.

Retroviruses: Retroviruses are a group of simple RNA viruses which are capable of transforming cells. Cellular transformation may result if the virus inactivates tumor-supressor genes or activates oncogenes (insertion mutagenesis) when it integrates into the host DNA.

Proto-Oncogenes and Oncogenes: One copy of an oncogene is capable of transforming a cell. Oncogenes are found in their active state in retroviruses and in transformed cells. Inactive oncogenes (proto-oncogenes) are found in non-transformed cells. Proto-oncogenes can be induced to transform a cell by mutation, by movement to a region with a strong promoter or enhancer, or by amplification.

Problem-Solving Hints

Hint 1: Eukaryotic genes, unlike prokaryotic genes, are not arranged in operons.

Hint 2: Embryonic development is controlled by maternal-effect genes, zygote genes, and environmental cues.

Hint 3: Inactive DNA is methylated while less methylation is observed in regions of active transcription.

Hint 4: Antibody diversity results from the combining of gene products and not directly from the genotype.

Hint 5: Mutation is not necessary to induce proto-oncogenes to transform cells.

Practice Problems

Problem 15-1. The FlavrSavr tomato is a genetically engineered tomato that has a longer shelf life than normal tomatoes. These tomatoes were created by using antisense RNA. Calgene scientists cloned the gene that produces a protein involved in fruit ripening and inserted it into the tomato backwards. How and why did this technique work?

Problem 15-2. Actinomycin D is an antibiotic that inhibits RNA synthesis. If this antibiotic is added to fertilized frog eggs, protein synthesis continues normally until the gastrula stage. From this point on, the embryo stops synthesizing protein. Why?

Problem 15-3. Considering only the ABO blood system in humans, why is type O blood the universal donor and type AB blood the universal recipient?

Problem 15-4. Consider a pair of identical twins with type B blood. Would both twins produce anti-A antibodies in their serum? If so, would their antibodies have the same polypeptide sequences?

Problem 15-5. Identify the source (mutated maternal-effect gene, segmentation gene, or homeotic gene) of each *Drosophila* mutant described below.

a. embryos are missing heads and thoracic structures

b. affected individuals are missing two abdominal segments

c. an individual has legs in place of antennae on the head

d. an individual has two thoracic segments

e. an individual lacks all abdominal segments

Problem 15-6. For how many cell divisions in human embryonic development are the cells of a zygote totipotent given that sets of identical quintuplets have been born?

General Questions

Question 15-1. Dolly, the cloned sheep, is not identical to her genetic mother. In what way(s) does she differ?

Solutions to Practice Problems

P15-1. Remember that antisense RNA is complementary to the mRNA normally produced by transcription of the gene. By inserting a reverse reading sequence, an RNA molecule that is complementary to the normal mRNA is produced. A high percentage of the mRNA molecules normally produced are expected to hybridize with the mRNA transcribed from the inserted sequence. A double-stranded RNA molecule cannot be translated at the ribosome. As a result, the protein is not produced. This technology slows down the ripening process by decreasing the number of mRNA molecules translated. Lower levels of the protein extends the shelf-life of the tomato.

P15-2. Actinomycin D inhibits the embryo's ability to synthesize RNA. Prior to gastrulation, the mRNA's existing in the cell are translated. Because they already exist in the cell at the time of treatment, early protein synthesis is unaffected.

P15-3. Blood type O produces no antigens and will not cause an immune response in the recipient. Type AB blood doesn't produce antibodies for either the A or B antigens. Therefore, no immune response is evoked when an AB individual receives A, B, or O type blood.

P15-4. Both of the twins would produce anti-A antibodies because they have type B blood. However, their antibodies may differ in sequence (*Hint 4*). Antibody diversity results from variations of which subunit genes combine and how they combine as well as from diversity at junctions and unusually high mutation rates. It would therefore be unlikely that the antibodies of both twins would have the same polypeptide sequence.

P15-5. Phenotypes (a) and (e) are likely caused by maternal-effect genes; phenotype (b) results from a mutated segmentation gene; phenotypes (c) and (d) result from homeotic gene mutants.

P15-6. At least three cell divisions must take place to produce a zygote that could give rise to five identical individuals.

Answers to General Questions

Q15-1. Dolly received, and was influenced by, factors and organelles found in the enucleated oocyte. For example, Dolly inherited her mitochondrial DNA from the egg donor and not from the nuclear DNA donor. Because of this difference, Dolly is not an exact duplicate of the ewe from which she was cloned. Identical twins are more similar to each other than Dolly is to her "genetic mother."

Answers to Even Numbered End of Chapter Exercises, Problems and Critical Thinking Questions

2. Parasegments develop first and then give way later in embryological development to true segments. True segments derive from the anterior part of one parasegment and the posterior part of the next parasegment in the poster direction.

4. The *hunchback* gene is active as a maternal-effect gene, causing a low level of *hunchback* protein throughout the fertilized egg. However, that level is not important because the anterior morphogen, Bicoid, enhances the expression of the *hunchback* gene and the posterior morphogen, Nanos, represses *hunchback* gene expression, resulting in a Hunchback protein gradient in the developing embryo. That gradient then affects later gene expression (see fig. 15.9).

6. It can be seen that homeotic genes have been conserved evolutionarily by the presence of homologous genes and gene sequences (homeobox) from yeast to human beings.

8. A morphogen is a diffusable substance in a developing embryo that controls gene expression depending on the concentration of the morphogen at that point in its gradient. The *bicoid* gene, expressed in maternal tissue, results in a region of the messenger RNA at the anterior tip of the embryo at fertilization. The messenger is then translated into a protein that diffuses into the embryo, creating a gradient that controls expression of other genes such as *hunchback*.

10. Assuming that development will take place at all, very early development should reflect the cytoplasm (frog) but soon be taken over by nuclear control (*Xenopus*). A *Xenopus* adult with *Xenopus* germ cells should result.

12. The fertilized egg, in the presence of the drug, will not make any RNA. For protein synthesis to occur, the egg must contain large quantities of stored RNA that is used during early cleavages. Development aborts at a time when newly synthesized mRNA is required.

14. Although the enzymes recognize the same sequence, only *Msp*I will cleave methylated and unmethylated DNA. The original fragment must contain two restriction sites that contain methylcytosine, since we see two extra fragments with *Msp*I.

16. The increase in RNA synthesis indicates that the virus is stimulating transcription of the oncogene; the question is how. Since promoters usually have an effect only on nearby or adjacent regions, and since the virus is 3 kb away, it is very unlikely that the virus directly affects the promoter. The virus probably has an enhancer that is stimulating transcription.

18. An antibody recognizes an antigen by "lock and key" complementary structure; see figure 15.26.

20. Any mature B lymphocyte should produce only one antibody; it has the potential to produce in excess of 10^9 different kinds.

22. B and T lymphocytes are produced in the bone marrow and mature either in the bone marrow (B cells) or in the thymus gland (T cells). B cells produce immunoglobulins; cytotoxic T cells kill infected host cells, and helper T cells are involved in communication with macrophages and B cells.

24. The major histocompatibility complex is a gene complex whose protein products are involved in immunity, including antigen presentation, intercell communication, and complement production.

26. Only one of the alleles seems to be active in any one given cell, a phenomenon referred to as *allelic exclusion.* Recall that in order for an antibody molecule to be synthesized, the V gene must be arranged so that it is adjacent to the C gene. These results suggest that the rearrangement occurs on only one of the two chromosomes of a cell. Once the rearrangement occurs, no more rearrangements are possible for that cell.

28. The gross chromosomal abnormalities associated with cancers are primarily deletions, inversions, and translocations.

30. Retinoblastoma is frequently associated with a deletion of chromosomal band 13q14. Wilm's tumor is associated with the deletion of chromosomal band 11p13.

32. The disease occurs only in homozygotes indicating that the normal condition is dominant (functional) and the retinoblastoma condition (nonfunctional) results from the failure of an enzymatic step.

34. Minimally, a retrovirus must have the *gag*, *pol*, and *env* genes. The protein products of *gag* are core viral proteins; of *pol* are reverse transcriptase, protease, and integrase; and of *env* are envelope glycoproteins.

36. See box 15.2, figure 3 for the mechanism by which long terminal repeats are generated at the ends of retroviral DNA; they encode an enhancer and promoter, among other signals.

38. Cellular oncogenes can be activated by mutation or enhanced transcription resulting from new placement of a gene by either deletion of nearby material or translocation.

40. Burkitt's lymphoma can arise when c-*myc* is translocated adjacent to the IgM constant gene promoter, which is very active in B lymphocytes. Hence c-*myc* is transcribed at a much greater rate than normal.

42. a. The virus could increase the amount of messenger RNA for the growth factor. b. Use northern blots to quantify the amount of oncogenic messenger RNA present in infected and uninfected cells. If a growth factor is overproduced, the excess molecules could cause cells to divide more rapidly and become cancerous. Many retroviruses contain regions that behave as enhancers of transcription of cellular oncogenes. If the virus inserts near an oncogene, the enhancer could stimulate transcription. An increase in transcription of a specific gene can be followed by quantifying the amount of specific RNA present in infected and uninfected cells. The intensity of the hybridized band should be higher in cells that are making more RNA.

Critical Thinking Question

2. As we discussed in chapter 5, there are numerous different sex determining mechanisms in plants, including species in which a single locus controls flower development to plants that have the equivalent of X and Y chromosomes. It seems that all flowers have the potential to be hermaphroditic including those that become male or female. We know this since at an early stage of development, all three flower types appear identical. That is, flower primordia have all four primordial whorls, including the stamen and carpel whorls. The simplest mechanism of sex determination would involve the repression of the development of the stamen whorls in female flowers and the repression of the carpel whorl in male flowers. Current research indicates that this repression of one whorl or another is probably involved in most flower sex determination and is controlled through genetic and hormonal mechanisms.

CHAPTER 16
DNA: ITS MUTATION, REPAIR, AND RECOMBINATION

Key Genetic Concepts

Mutation: The term **mutation** refers to the process by which a gene changes structurally. An individual which expresses the effects of this change is called a **mutant**. Nucleotides are the smallest units of mutation. For every mutational change in the DNA, there may be a corresponding change in the protein product of the mutant gene. This is described as **colinearity**. It is important to realize that not all mutations produce an altered phenotype and that more than one mutation can produce the same phenotypic effect. Geneticists can determine if two mutations are alleles of the same gene by creating a heterozygote and noting its phenotype (**complementation test**). A mutant phenotype indicates that the mutations occurred in the same gene. Examination of the offspring of this heterozygote suggests whether or not the mutations are structural, as well as functional, alleles. If the wild-type phenotype occurs at the background mutation rate, the mutations occur at the same place on the **cistron** (the smallest unit coding for a single polypeptide). Alleles that are functional but not structural are termed **pseudoalleles**.

Types of Mutation: Mutations that consist of single changes in the nucleotide sequence are point mutations. For example, a purine or pyrimidine can be replaced by another base of the same class (a **transition**). Base analogs such as 5-bromouracil or 2-aminopurine produce transitions. If a purine is replaced with a pyrimidine or vice versa, a **transversion** is said to occur. Tranversions are caused by chemicals such as ethyl methane sulfonate and ethyl ethane sulfonate. Point mutations that add or delete bases cause a shift in the reading frame (a **frameshift** mutation) thus potentially altering the amino acid sequence of a protein encoded for by a mutated gene. Subsequent mutations can restore gene function (**intragenic suppression**) or the original gene sequence (**back mutations**). As discussed in Chapter 7, mutations can be conditional-lethal. **Nonsense** mutations result in the formation of an inappropriate stop codon while a **missense** mutation changes an amino acid specified in the code.

DNA Repair: Cells normally have the ability to correct mistakes found in the DNA molecule. Loss of this ability may result in mutations, cell death (**apotosis**), or cancer. **DNA photolase**, in the presence of light, breaks the linkages of adjacent pyrimidines (**dimers**) caused by exposure to UV light. Damaged portions of DNA are removed by the general mechanism of excision repair. Enzymes remove damaged bases in the process of **base excision repair**. Dimers may be removed in the process of **nucleotide excision** repair. **Mismatch repair** removes mismatched bases and is the most common type of DNA repair. The **SOS response** is induced by the presence of single-stranded DNA resulting from postreplicative gaps.

Recombination: The **Holliday model** of recombination involves simultaneous breaks in one strand of each helix involved in a crossover. In bacterial recombination, a linear molecule recombines with a circular molecule. **Heteroduplex DNA**, DNA with loops and bubbles in regions where two DNAs differ, results from recombination.

Problem-Solving Hints

Hint 1: When determining the base sequence of a mRNA molecule given the amino acid sequences of the normal and mutant protein, write down all of the possible codons corresponding to each amino acid. To distinguish between alternate bases in the third position of the codon, compare the normal sequence with each mutant sequence.

Hint 2: When determining the nature of a base change, consider the fewest changes in the normal mRNA sequence necessary to produce the mutant sequence.

Hint 3: The number of altered amino acids in a mutant protein suggests the type of mutation that has occurred. A missense mutation generally produces an incorrect amino acid while a frameshift mutation produces a protein with several to many incorrect amino acids. A protein with an amino acid sequence that is correct but terminates prematurely suggest the occurrence of a nonsense mutation.

Hint 4: Mutations may not alter the phenotype. Mutations can restore the original function and/or sequence of an mRNA or mask the occurrence of the original mutation.

Hint 5: Mutations can be inherited or they may occur spontaneously.

Hint 6: To verify allelism, examine the phenotype of a heterozygote for both mutations. If wild-type, the mutations are not alleles of the same gene. If mutant, the mutations are presumed to be functional alleles.

Hint 7: Examination of the F2 offspring of the heterozygotes in *Hint 6* will indicate if the alleles are structural: the genes are presumably structural if the occurrence of wild-type offspring is at the background mutation rate.

Practice Problems

Problem 16-1. Sickle-cell anemia results from an abnormal hemoglobin molecule which causes the red blood cells of homozygotes to sickle. In sickle-cell hemoglobin, a single amino acid, glutamic acid (glu), is replaced by valine (val). Can you identify the change that probably occurred?

Problem 16-2. Consider the mRNA sequences from a hypothetical organism given below. Identify each mutational type illustrated and provide the amino aid sequence corresponding to each mutant mRNA.

Normal: AUGUUCAGCUGGGGGGUGUAA

Mutant 1: AUGUUCAGCUGAGGGGUGUAA

Mutant 2: AUGUUCAGCUGGCGGGUGUAA

Mutant 3: AUGGUUCAGCUGGGGGGUGUAA

Problem 16-3. You have isolated a new UV-sensitive mutant strain of bacteria. A number of colonies from the mutant strain and a wild-type strain were cultured. Individual plates of each strain were exposed to the same dosage of UV radiation and then assigned to one of two treatments: light incubated or dark incubated. The percentage survival for each is given below. Explain the differences observed between the mutant and wild-type strains and the light and dark treatments.

	UV Dosage Results:	
Wild-type, light		85%
Mutant, light	80%	
Wild-type, dark		60%
Mutant, dark	9%	

Problem 16-4. In *Drosophila,* two recessive mutations affecting eye-color, white and cherry, are sex-linked. Given the results below, determine if the two mutations are located in the same or in different genes.

Cross: white-eyed female x cherry-eyed male

Offspring: white-eyed males and light cherry-eyed females

General Questions

Question 16-1. Consider a gene A in human beings. How can mutations of this gene cause several different genetic disorders or no disorder at all?

Question 16-2. Explain the following observation: RNA viruses accumulate mutations at a faster rate than DNA viruses.

Question 16-3. High dosages of x-rays produce dominant lethals but do not affect fertility in a number of insect pest species. Individuals possessing one copy of a dominant lethal are not viable. If the female of a pest species mates only once, how can the relationship between x-rays, mutation, and fertility be used to design an eradication program?

Solutions to Practice Problems

P16-1. The codons corresponding to glu are GAA and GAG. The codons corresponding to val are GUU, GUC, GUA, and GUG. When determining the nature of a base change, consider the fewest changes in the normal mRNA sequence necessary to produce the mutant sequence (*Hint 2*). In the sickle-cell hemoglobin, a transition probably occurred: the adenine (GAG or GAA) was probably replaced by uracil (GUG or GUA) resulting in the codon for valine.

P16-2. To determine what type of mutation has occurred we must compare the mutant mRNA sequence with the normal mRNA sequence (*Hint 1-3*). The consequences of this change will be revealed upon examination of the amino acid sequence coded for by the mutant mRNAs (refer to Hint 3 above).

Normal:	AUG	UUC	AGC	UGG	GGG	GUG	UAA	
	meth	phe	ser	try	gly	val		
Mutant 1:	AUG	UUC	AGC	**UGA**	GGG	GUG	UAA	
	meth	phe	ser					

NONSENSE MUTATION

Mutant 2:	AUG	UUC	AGC	UGG	**CGG**	GUG	UAA	
	meth	phe	ser	try	**arg**	val		

MISSENSE MUTATION

Mutant 3:	AUG	**GUU**	CAG	CUG	GGG	GGU	GUA	A
	meth	**val**	**gln**	**leu**	**gly**	**gly**	**val**	. . .

FRAMESHIFT MUTATION (Addition)

P16-3. It appears that the mutant strain has a defective base and/or nucleotide excision repair system. Notice that both the wild-type and mutant strains have a high survival rate in the presence of light. This indicates that the mutant strain, like the wild-type, has an active DNA photolase enzyme which repairs dimers induced by UV when light energy is available. In the dark, the wild-type survival is reduced but is still high because an alternative repair mechanism can be used. A mutation in this system would account for the very low survival of the mutant strain incubated in the dark.

P16-4. To determine if two mutations are in the same gene we must examine the phenotype of the heterozygote. In the case of sex linkage, females are heterozygous. The females resulting from this cross are mutants indicating that the two mutations are in the same gene (*Hint 6*).

Answers to General Questions

Q16-1. Not all mutations result in the production of an abnormal protein (*Hint 4* for example). Therefore, not all mutations lead to the expression of a disorder. Several disorders may arise from mutations in the same gene provided each alters a different amino acid(s).

Q16-2. RNA viruses must undergo reverse transcription to form DNA which is then used as a template to synthesize the RNA genome. This extra step in the process of reproduction provides an additional opportunity for mistakes to be made in the fidelity of replication.

Q16-3. Males of the pest species could be exposed to high levels of radiation and then release in large numbers. Because fertility is unaffected, these males could presumably mate normally. However, the offspring they produce will show marked decreases in viability. If a large population can be released, females will be more likely to first encounter and mate with a treated male. This approach will only work for a species where the females mate only once but males mate with several females. In addition, the treated males must be as attractive as the wild-type males occurring in the field naturally.

Answers to Even Numbered End of Chapter Exercises, Problems and Critical Thinking Questions

2. Temperature-sensitive mutations are most valuable to study critical functions of a cell because they display the mutant phenotype only at the restrictive temperature. At the permissive temperature, the cells can be kept alive. For example, temperature-sensitive mutations of enzymes of DNA replication can be studied, whereas many of the mutants, if not temperature sensitive, would die.

4. The overlap patterns of the deletions define six regions: overlap of 1 and 2; 2 only; overlap of 2 and 3; 3 only; overlap of 3 and 4; 4 only. We can thus localize mutations M to S according to these six areas:

Mutation M: region of 3 only.

Mutation N: region of 2 and 3 overlap.

Mutation O: region of 4 only.

Mutation P: region of 3 and 4 overlap.

Mutation Q: region of 2 only.

Mutation R: region of 1 and 2 overlap.

Mutation S: region of all but unique and overlap areas of 4.

The figure shows where each of these mutations are in the *A* cistron.

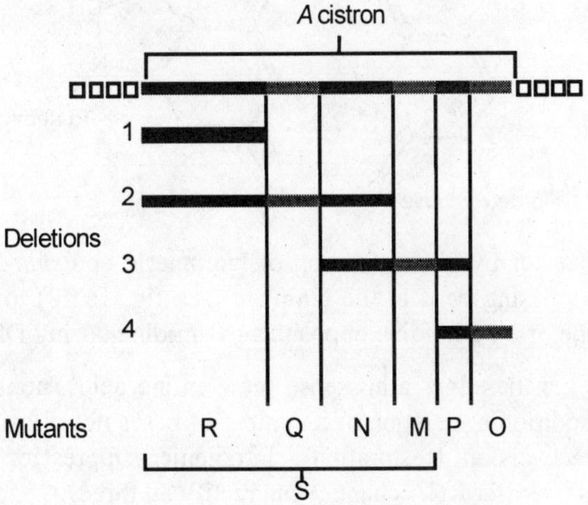

6. When organisms that have phenotypes controlled by functional alleles are mated, their progeny, for the most part, are mutant; the alleles do not complement each other. Structural alleles are functional alleles that have changes at identical nucleotides. In the *cis* configuration, one cistron has two mutant sites.

8. In intra-allelic complementation, mutant polypeptides form a functional enzyme or protein. This gives the impression of genic complementation, which would lead the researcher to believe the mutations were in nonallelic genes.

10. See figure 16.14 for the tautomeric forms of the bases. Figure 16.15 shows one base pair in which adenine is in the imino form and therefore can pair with cytosine; we diagram the remaining base pairs here. In a similar fashion, table 16.4 presents all tautomeric base pairings. These result in transition mutations: AT becomes GC; GC becomes AT; TA becomes CG; and CG becomes TA.

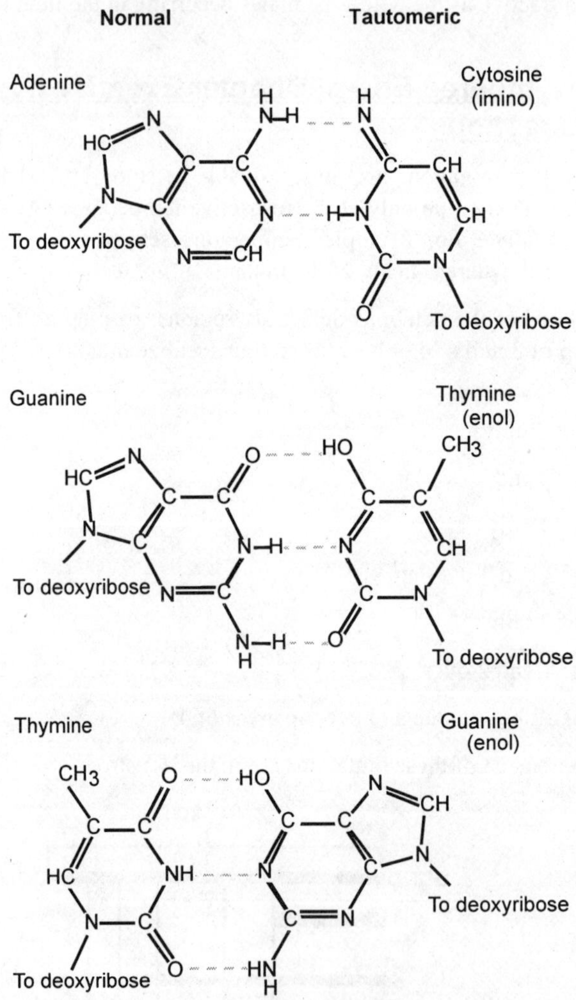

12. Transversion can occur by a combination of tautomeric and *anti-syn* shifts (see fig. 16.18), by replication opposite a missing base in the template (see fig. 16.22), or by direct incorporation of a purine opposite a purine or a pyrimidine opposite a pyrimidine during DNA replication.

14. A point mutation can be either a missense (new amino acid), nonsense (stop codon created), or frameshift mutation (addition or deletion of a number of bases not divisible by three). These can all be corrected by simple reversion (restoration). Intragenic suppression can rectify the frameshift. Intergenic suppression (transfer RNA change) can rectify all three.

16. Intergenic suppression is restoration of function of a mutated gene by a mutation at another locus. Intragenic suppression is restoration by a second mutation at the same locus. Both involve a second mutation event at a different position from the first. Neither usually restores the wild-type completely.

18. To do any complementation test, you must make a heterozygote, so first cross homozygous *Chx-1* × homozygous *Chx-2*. Now cross the heterozygote to a drug-sensitive, wild-type organism. The crosses and expectations appear below.

If allelic:	*If nonallelic, nonlinked:*	*If nonallelic, linked:*
Chx-1/*Chx-2* × +/+	$\frac{Chx\text{-}1\ \ Chx\text{-}2}{+\ \ \ \ +} \times \frac{+\ +}{+\ +}$	$\frac{Chx\text{-}1\ \ \ \ +}{+\ \ \ \ Chx\text{-}2} \times \frac{+\ +}{+\ +}$
1 *Chx-1*/+:1 *Chx-2*/+	1 *Chx-1*/+ +/+	
all resistant	1 *Chx-2*/+ +/+	some wild-type recombinants:
	1 *Chx-1*/+ *Chx-2*/+	<
	1 +/+ +/+	25% sensitive
	3 resistant: 1 sensitive	

Thus, if we find any wild-type in the testcross, the mutations are nonallelic. If these two mutations are allelic but at different sites (functional but not structural alleles), we might expect to see a very rare wild-type as a result of recombination, but this frequency will almost certainly be much less than 1%.

20. No. The nonsense mutations are UAG, UGA, and UAA. Their respective DNA sequences are:

$$\begin{matrix} ATC, & ACT, & & ATT \\ & & \text{and} & \\ TAG & TGA & & TAA \end{matrix}$$

Remember that mutations occur at the DNA level. Any C → T change produces

$$\begin{matrix} ATT, \\ TAA, \end{matrix}$$

another nonsense codon.

22. To begin either of these processes, ask yourself what properties the mutant would have, then devise a scheme that utilizes this property. A large bacterial cell would be expected to be heavier, and might settle to the bottom more quickly. Place a bacterial population in a long column, wait a few minutes, and collect what is at the bottom; repeat a few more times. Or, use a filter that will allow normal, but not large, bacteria to pass through. The desired bacteria will be trapped on the filter. Nonmotile ciliates might be expected to sink, so an approach similar to the first described for large bacteria might work.

24. Intragenic complementation, when the functional unit of a protein is composed of more than one polypeptide. Occasionally, mutations in different parts of a protein will compensate for, or correct, the structural change brought about by the first mutation. See text fig. 16.9. We can show intragenic complementation as:

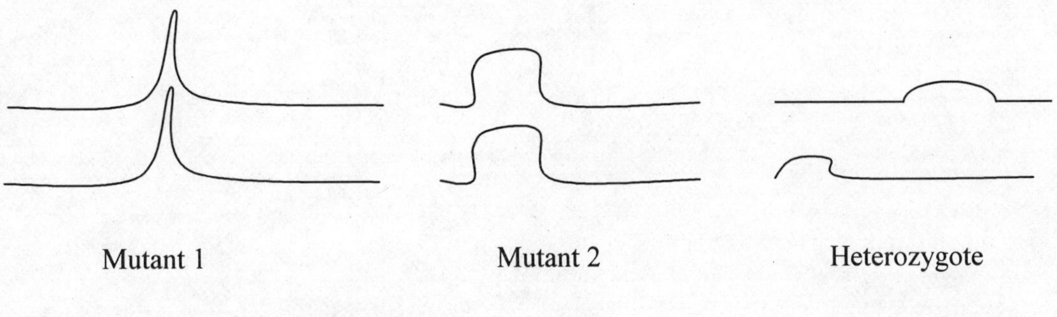

We often do not get 100% activity with intragenic complementation, but usually the amount of activity is sufficient for growth.

26. The *lac*$^-$ mutations may be physiologically induced. In table 16.1, we see a wide variation among individual cultures, indicating that mutations were produced randomly. The results in this problem suggest that the environment is inducing the mutations.

28. Thymine dimerization can be restored by photoreactivation (DNA photolyase required), excision repair (see fig. 16.28), or postreplicative repair (see fig. 16.34). The enzymes are listed in table 16.5.

30. RecA causes single-stranded DNA to invade double-stranded DNA. It also acts as a protease, cleaving the λ cI repressor protein (and is involved in the catalysis of LexA). Cleaving the cI repressor causes λ to enter the lytic cycle.

32. Reciprocal recombination is shown in figure 16.36; nonreciprocal in figure 16.39.

Critical Thinking Question

2. DNA is an incredible molecule: with a mere four subunits, it not only contains all the information for the development of a human being, but also can be repaired even after numerous kinds of damage. The essence of the elegance of the molecule is its double helical configuration that has a redundant (complementary) copy of the information. Thus replication is a simple template driven process and repair, for the most part, is simple because of the redundant undamaged strand found under most damage conditions. Even if both strands of a double helix are damaged or broken, there is often the opportunity to use a homologous double helix, as in meiotic or bacterial recombination or postreplicative repair. It is hard to imagine a molecular structure that could hold as much information and yet be as easy to replicate and repair.

CHAPTER 17
NON-MENDELIAN INHERITANCE

Key Genetic Concepts

Non-Mendelian Inheritance: Most of your textbook discussions focus on the inheritance of traits influenced by chromosomal genes. There are a number of genetically controlled traits, however, which are not associated with the nuclear DNA. Some of these traits are influenced by the genotype of one parent, usually the mother. Other traits are controlled by nonnuclear genes found in the chloroplasts, mitochondria, infective agents, or plasmids.

Parental Influences: As mentioned above, some aspects of an individual's phenotype may be determined by the genotype of the maternal parent. These traits are described as being **maternally influenced**. Parental influences may also lead to the differential expression of alleles depending on their parental origin. This phenomenon is referred to as **imprinting**.

Cytoplasmic Inheritance: The mitochondrion and chloroplast contain DNA that codes for a number of proteins. Both of these genomes are extranuclear and capable of replicating separately from the nuclear genome. Chloroplasts and mitochondria are not synthesized *de novo* and are passed to a new individual intact. In general, maternal inheritance is observed when mitochondria and/or chloroplasts are partitioned into the egg cell during cytokinesis. Plasmids in prokaryotes, also represent extrachromosomal genetic systems. These segments of DNA are highly mobile between bacterial cells as discussed in Chapter 7.

Problem-Solving Hints

Hint 1: Extranuclear inheritance is suggested if the results of reciprocal crosses produces progeny with the phenotype of only one parent.

Hint 2: Nuclear transplantation does not affect extrachromosomal inheritance. Cytoplasmic inheritance usually involves organelles, symbionts, or parasites that have their own genetic material.

Hint 3: The phenotypes of F1 and F2 progeny will be the same as those of the female parent if the trait is maternally influenced *or* maternally inherited. Several generations must be examined to conclusively conclude that a trait is maternally inherited (see *Hints 4 and 5* below).

Hint 4: If a trait is maternally *influenced* (maternal effects), the progeny phenotype is dependent upon the mother's genotype. Reciprocal crosses yield different results and the occurrence of a 3:1 ratio in the offspring is delayed by one generation (3:1 is found in the F3 generation rather than the F2).

Hint 5: If a trait is maternally *inherited*, genetic transmission is only through the maternal cytoplasm. The male makes no contribution to the progeny phenotype. Regardless of the number of generations examined, offspring will have the same phenotype as their mother if the trait is maternally inherited.

Hint 6: Imprinting is suggested if the expression of specific genes is not independent of parental origin.

Practice Problems

Problem 17-1. In *Drosophila*, sensitivity or resistance to carbon dioxide displays maternal inheritance. Predict the outcomes of reciprocal crosses involving matings between sensitive and resistant flies.

Problem 17-2. Three individual crosses were made between killer and sensitive *Paramecium*. Examine the results of each cross and use this information to assign genotypes to the parental strains. Cross 1: all killer; Cross 2: 75% killer and 25% sensitive; and Cross 3: 50% killer and 50% sensitive.

Problem 17-3. A male from a true-breeding strain of nervous zeebogs is crossed to a calm female, also from a true-breeding strain. The F1 and F2 offspring are all calm. Propose an explanation for these findings. If the F2 generation information were unavailable, what conclusions would you have drawn?

Problem 17-4. Complete the pedigree below such that maternal inheritance is suggested and dominant sex-linked inheritance can be excluded.

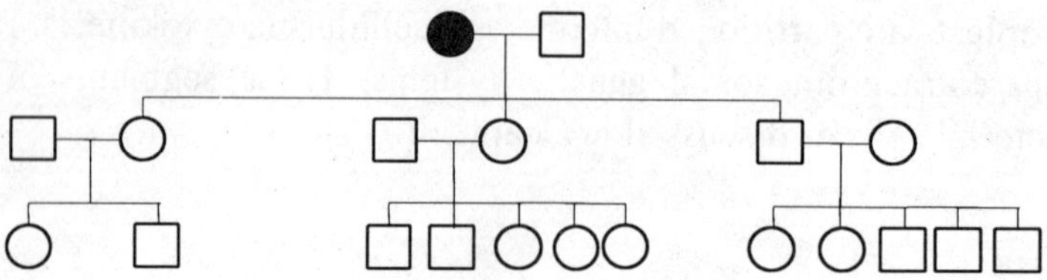

Problem 17-5. Examine the pedigree below and determine how the trait is likely inherited.

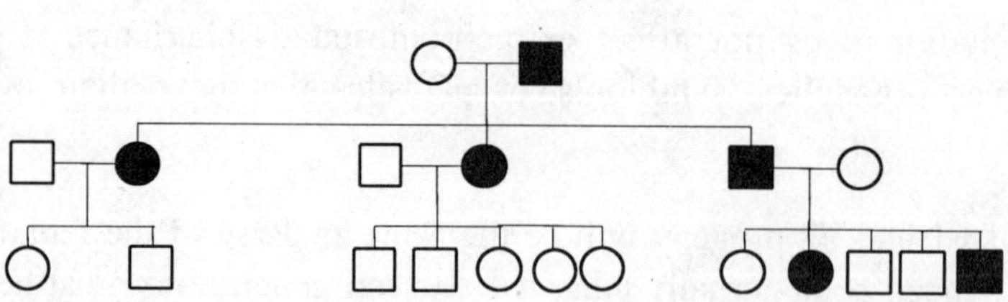

Problem 17-6. In four-o'clock plants, leaves can be green, white, or variegated. Flowers from branches of each type were used to perform all possible crosses (outlined below). Predict the progeny phenotypes expected if (a) inheritance is biparental (codominance) or (b) inheritance is maternal.

	Female Parent	Male Parent		Female Parent	Male Parent
1	white	white	6	green	variegated
2	white	green	7	variegated	white
3	white	variegated	8	variegated	green
4	green	white	9	variegated	variegated
5	green	green			

General Questions

Question 17-1. Jesse James, the notorious bank robber, was allegedly killed in 1882. Since that time, rumors that the dead man was not Jesse James circulated widely and were supported by several men who claimed to be the dead "legend." Recently, a group of scientists exhumed the body of Jesse James in an effort to confirm his death. The research team obtained DNA from James' tooth pulp. They also obtained DNA samples from the descendants of James' sister. Why were these samples chosen to confirm the identity of Jesse James?

Solutions to Practice Problems

P17-1. When a trait is inherited maternally, genetic transmission is only through the maternal cytoplasm (*Hints 1 and 5*). If we crossed sensitive females with resistant males, we would expect all of the offspring to be sensitive. If we cross resistant females with sensitive males, we would expect all of the offspring to be resistant.

P17-2. In cross one, all of the offspring display the killer phenotype suggesting that both parents also had the killer phenotype. Genotypes would be KK x KK or KK x Kk. The 3:1 ratio of killer to sensitive phenotypes in cross 2 suggests that both individuals were heterozygous killer (Kk x Kk). In cross 3, crossing a heterozygous killer and a sensitive strain (kk) would result in the observed 1:1 ratio.

P17-3. In this cross, we see that the F1 and F2 phenotypes are the same as the P1 female which suggest maternal inheritance. If we would have looked at the F1 phenotypes only we probably would have concluded that the trait was likely inherited as an autosomal with calm dominant to nervous. This problem illustrates that it is important to examine multiple generations resulting from a cross before drawing conclusions (*Hint 3*).

P17-4. In maternally inherited traits, all individuals have the phenotype of the female parent. In generation III, we can distinguish maternal inheritance from sex-linked dominant inheritance by examining the offspring of affected males and females. For example, approximately one-half of the daughters of the affected male would be affected if the trait were sex linked dominant. No affected daughters will result from the affected father if the trait is maternal.

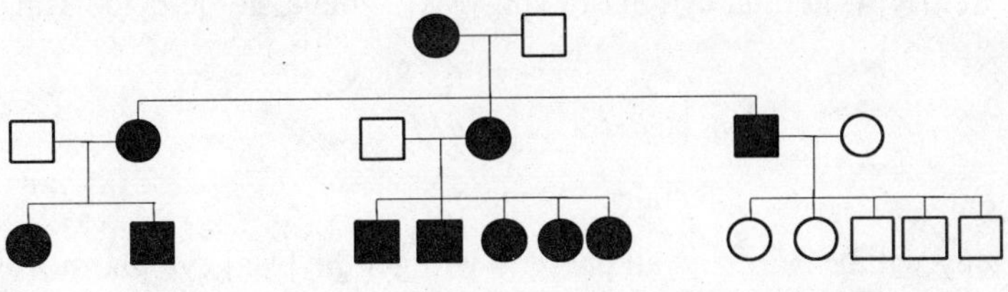

P17-5. The pedigree is probably showing paternal influence or paternal inheritance. Because affected males produce both affected daughters and affected sons, we can rule out Y-linked inheritance as a likely explanation. Sex linked inheritance is eliminated because none of the affected females have affected children.

P17-6. Cells in variegated regions of the plant contain both green and white chloroplasts. Depending on how the chloroplasts are partitioned in the egg, a variegated plant can give rise to green, white, or variegated offspring phenotypes.

Cross (Female x Male)	(a) Codominance	(b) Maternal Inheritance
1. white x white	white	white
2. white x green	variegated	white
3. white x variegated	1 variegated : 1 white	white
4. green x white	variegated	green
5. green x green	green	green
6. green x variegated	1 variegated : 1 green	green
7. variegated x white	1 variegated : 1 white	green, white, variegated
8. variegated x green	1 variegated : 1 green	green, white, variegated
9. variegated x variegated	1 green : 2 variegated : 1 white	green, white, variegated

Answers to General Questions

Q17-1. Mitochondrial DNA was used to identify the remains of Jesse James. While James and his sister shared only 50% (on average) of their nuclear genes, they both received their mitochondrial DNA from their mother. James' sister in turn passed her mitochondrial DNA to her children and so on. With the exception of mutation, the mitochondrial DNA sequence of the descendants of James' sister should be identical to that of James. The identity of the remains of Jesse James was finally confirmed by mitochondrial DNA analysis.

Answers to Even Numbered End of Chapter Exercises, Problems and Critical Thinking Questions

2. Several bits of evidence lead to the conclusion that a trait is controlled by a plasmid gene: existence of a plasmid that can be seen by electron microscopy or through density-gradient centrifugation; a change in multiple aspects of the phenotype; infectious nature of the change; lack of linkage to host chromosomal loci; and disappearance of the trait with disappearance of the plasmid.

4. Sex linkage shows patterns of inheritance (e.g., zigzag) that differ among sexes in reciprocal crosses. These patterns are not seen in snails. Also, snails are hermaphroditic: each individual can be both male and female.

6. A sinistral female must have had a *dd* mother and therefore must have at least one *d* allele. If she produces dextral offspring, she must have a *D* allele. Therefore, her genotype must be *Dd*; her father provided her *D* allele and therefore could have been *DD* or *Dd*.

8. *Rf/rf,* male fertile. Let [N] = normal cytoplasm and [Ms] = male sterile cytoplasm. The cross is then:

$$[\text{Ms}], rf/rf \times [\text{N}], Rf/Rf$$

female male

Genotypically, the progeny will be *Rf/rf,* and all progeny will get the [Ms] cytoplasm from the female plant. Normally, all progeny should be male sterile, but in this case, *Rf* overrides this phenotype.

10. In matings, the petite phenotype follows Mendelian ratios without exception.

12. Cytoplasmic inheritance is determined by the different results seen in reciprocal crosses (see fig. 17.9). An *iojap*-type chromosomal gene will have a phenotype of inducing variegation (in Mendelian ratios dependent on the exact crosses) that is then inherited cytoplasmically.

14. Both mitochondria and plastids contain DNA and ribosomes with prokaryotic affinities. Defects in either can produce similar phenotypic effects of reduced growth.

16. Both strains have the *Kk* genotype; of sensitive exconjugants, all will produce sensitive (*kk*) offspring by autogamy. Of killer exconjugants, one-third (*KK*) will produce only killer offspring while two-thirds (*Kk*) will segregate both killers (*KK*) and sensitives (*kk*) through autogamy.

18. To begin with, controlled breeding experiments would be carried out over several generations to ascertain that normal Mendelian ratios among offspring did not occur. Once non-Mendelian inheritance is confirmed, the crosses could be examined for maternal effects. Given that these exist, an infectious agent could be looked for by injecting cytoplasm from infected (CO_2-sensitive) organisms into uninfected (CO_2-resistant) eggs. The CO_2-resistant organisms would then become infected. The infected cytoplasm could be searched for a bacterium or virus. The causative agent, known as the sigma virus, would be discovered. When isolated and purified, it would prove to be the infective agent.

20. Mutant female × normal male—all mutant progeny; normal female × mutant male—all normal progeny. Since the trait is carried by the mitochondrial genome, it will be passed to the next generation from the female parent to all progeny.

22. Assuming sporulation immediately after mating, asci will yield spores that are mostly resistant, although some will be sensitive, based on the distribution of mitochondria into the four spores.

24. 1 *poky*:1 *fast-poky.* The cross is *fast-poky* female × normal male. Assigning genotypes, we get *fast-poky: F* (*poky*); *poky: F′* (*poky*); normal: *F′* (+), where () = cytoplasmic component. Since the maternal parent is *fast-poky,* this individual will contribute *poky* cytoplasm to all offspring. Thus, all offspring will be either *poky* or *fast-poky.* The nuclear gene *F/F′* segregates normally to yield 4 *F*:4 *F′* .*F′* + *poky* cytoplasm = *poky,* and *F* + *poky* cytoplasm = *fast-poky.*

26. Three nonmate-killers:one mate-killer. The maintenance of mate-killer requires the presence of one of two dominant nuclear genes (M_1 or M_2). The double heterozygote is $M_1m_1M_2m_2$. Autogamy produces four different homozygous lines in equal frequency:

$M_1M_1M_2M_2$	mate-killer maintained
$M_1M_1m_2m_2$	mate-killer maintained
$m_1m_1M_2M_2$	mate-killer maintained
$m_1m_1m_2m_2$	no mate-killer

Critical Thinking Question

2. The lysogenic state of *E. coli* is an example of a different type of inheritance, termed epigenetic. It refers to the inheritance of a trait when no change in the DNA has taken place. Certainly lysogenic as well as lysed bacteria can have the same DNA sequence; the behavior of the phage is environmentally controlled. However, unlike Lamarkian inheritance, we have a comprehensible mechanism that involves known molecular events. Also, many scientists don’t believe that this sort of phenomenon is of great general occurrence.

CHAPTER 18
QUANTITATIVE INHERITANCE

Key Genetic Concepts

Polygenic Inheritance: In this chapter, we turn our attention to traits that vary continuously. These traits are usually influenced by several pairs of alleles, or **polygenes,** and by the environment. This type of inheritance is also described as **additive** because each of the alleles "add to" the final phenotype expressed by an individual. There are two types of alleles in the polygenic model: contributing and noncontributing. **Contributing alleles** (represented by capital letters) add a certain, fixed amount to the phenotype. Each contributing allele is assumed to have the same effect on the phenotype. In contrast, **noncontributing alleles** do not add units to the phenotype. For example, consider height in a hypothetical plant. Assume that this polygenic trait is influenced by four pairs of alleles. Plants with all contributing alleles (AABBCCDD) will be the tallest (20 cm) while those with all noncontributing alleles (aabbccdd) will be the shortest (4 cm). In this example, there is a characteristic "minimum" of 4 cm to which each contributing allele adds an additional 2 cm of length.

Estimating the Number of Alleles and Phenotypes: The number of polygenes can be determined by obtaining an F2 generation from a cross of true-breeding individuals with extreme phenotypes. Specifically, the number of F2 progeny with a phenotype as extreme as one of the P1 parents equals ($1/4^n$) pairs of alleles. For example, an F2 generation with approximately 8 out of 500 individuals (0.016) resembling one of the P1 parents closely approximates the value of $1/4^3$ (1/64 = 0.0156) indicating that six alleles influence this trait. The binomial expression can be used to determine the distribution pattern of phenotypic classes expected in an F2 generation. In the above example, the number and ratio of phenotypic classes can be obtained by expanding the binomial $(p + q)^6$. In this case, there will be seven phenotypic classes in the ratio 1 : 6 : 15 : 20 : 15 : 6 : 1.

Population Statistics: More complex methods of analysis are necessary to describe and understand polygenic inheritance. Important statistics include the **mean**, **variance**, **standard deviation**, and the **standard error** of the mean. Each of these statistics can be used to describe the distribution of individuals in a population and to compare the distributions of different populations. **Covariance**, **correlation**, and **regression** are used to determine if there is a relationship between two characteristics in a series of individuals.

Variation: If we breakdown the phenotypic variation of a population into its component parts, we find that both genes and the environment influence the expression of a polygenic trait. The diagram below illustrates the many factors that contribute to phenotypic variation. Notice that variation within and between individuals influences the observed phenotypic variation in the population. Within families, there are two main types of genetic variation: additive and nonadditive. **Additive factors** are passed from parents to offspring and are responsible for similarities between relatives. Nonadditive factors such as dominance and epistasis are genotype specific and are therefore not inherited. **Maternal effects** (which were covered in Chapter 17) also contribute to phenotypic variation.

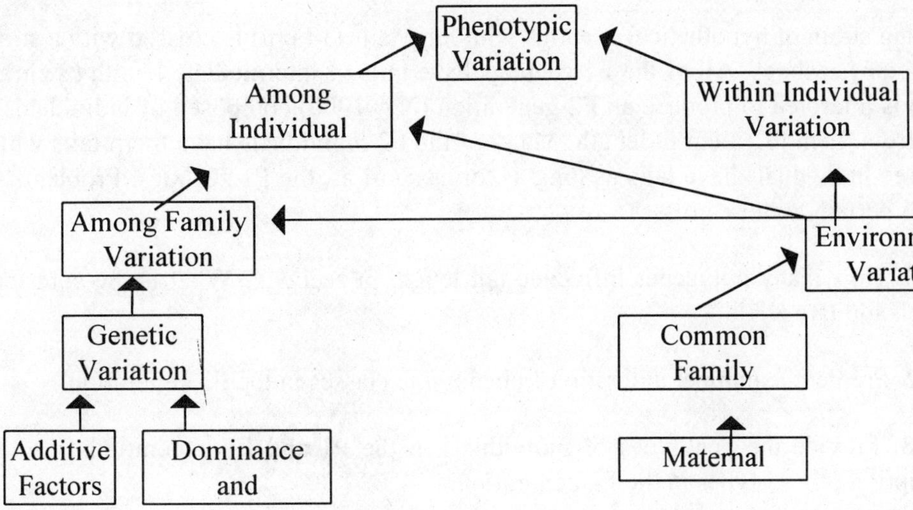

Heritability: Individuals within a population often vary in the characteristics they possess. For years, geneticists have taken advantage of this natural variation to create varieties of organisms, such as domestic animals and corn plants, that possess characteristics of interest (such as milk production in domestic animals or insect resistance in corn). In order for the characteristics of a population to change, the trait must have a genetic basis. **Heritability** is a statistic that allows a geneticist to determine if the observed differences in the population are due to genetic differences among the individual members of the population or due to the influence of the environment. A heritability of one implies that all of the variation observed in a population is due to genetic factors while a heritability of zero implies that the variation observed is due to environmental influences. Remember that we are describing the percentage of *phenotypic variation* observed in a *population* when we calculate a heritability *estimate*. **Broad-sense** heritability describes the proportion of the observed variation that is due to all genetic factors while **narrow-sense** heritability estimates the proportion of the observed variation that is due to additive genetic effects only.

Problem-Solving Hints

Hint 1: Traits that vary continuously are likely influenced by many loci and by the environment.

Hint 2: The number of polygenes influencing a trait can be estimated by determining the fraction of an F2 generation that express one of the parental (P1) phenotypes.

Hint 3: The number and ratio of phenotypic classes expected in an F2 generation for a polygenic trait can be obtained from the coefficients of the expanded binomial expression (see Chapter 4).

Hint 4: Each contributing allele (capital letter) adds a set amount to the phenotype.

Hint 5: Heritability describes the proportion of the phenotypic variation in a population that is attributable to genetic factors. It is a population statistic and *does not* describe the percentage of an individual's phenotype that is genetic.

Practice Problems

A pure-breeding strain of hypothetical zeebogs with long tails (14 cm) is crossed with a strain of short-tailed (2 cm) zeebogs. All of the F1 zeebogs have tails of intermediate length (8 cm). The F1 generation is interbred to produce an F2 generation (N = 1000) composed of individuals with tails ranging from 2 cm to 14 cm in length. Many of the F2 individuals have 8 cm tails while an average of three individuals have tails as long as, or as short as, the P1 zeebogs. Problems 18-1, 18-2, and 18-3 pertain to these crosses.

Problem 18-1. How many polygenes influence tail length in zeebogs? What is the contribution (in cm) of each additive allele?

Problem 18-2. Predict the number and ratio of phenotypic classes in the F2 generation.

Problem 18-3. Provide the genotypes of individuals in the P1 and F1 generations. Provide a general description of genotypes in the F2 generation.

Problem 18-4. Consider the set of data below. Calculate the correlation coefficient.

#1	#2	#1	#2	#1	#2
8	5	3	5	6	4
10	6	8	7	5	7
5	4	4	7	6	7
8	10	9	4	8	9

Problem 18-5. Assume that individuals #1 and #2 in problem 18-4 are full-siblings. Use the correlation coefficient calculated above to estimate the narrow-sense heritability for the trait. Interpret your calculated heritability estimate. What are the limitations of the calculation?

Problem 18-6. An Easter bunny farmer wants to increase the chocolate content of his Easter bunnies. In his population, the heritability of chocolate content is 0.69. The farmer selectively breeds bunnies with high chocolate content: the selected bunnies have, on average, 12 mg more chocolate than the unselected bunnies. What gain in chocolate content is expected in the next chocolate bunny generation?

Problem 18-7. Assume that height in humans is controlled by six polygenes with no environmental modification. In this model, each additive allele contributes four inches to a base height of four feet. What is the range of heights possible in this population? If a woman with the minimal height specified marries a man of intermediate height, what is the maximum height a child they produce could achieve.

Problem 18-8. What are the expected correlations between the following pairs of relatives: a woman and her (a) mother; (b) brother; (c) husband; (d) grandmother; (e) uncle; and (f) first cousin? If this woman's mother has a monozygotic twin, what is the expected correlation between the woman and the twin aunt's child (g)?

General Questions

Question 18-1. Compose a list of traits in human beings that are likely to be inherited quantitatively.

Question 18-2. Virtually all human beings are born with two eyes: there is little to no phenotypic variation in this population. The heritability of eye number is zero. Explain.

Solutions to Practice Problems

P18-1. To estimate the number of polygenes influencing a trait, determine the average number of F2 individuals that have the parental phenotype (*Hint 2*). In this example, an average of 3/1000 individuals have 2 cm or 14 cm tails, like the P1 zeebogs. This value approximates the value expected if four pairs of alleles are involved ($1/4^n$ = 1/256). If a total of eight polygenes are involved, each contributing allele adds [(14-2) / 8] or 1.5 cm of length.

P18-2. The relevant binomial is $(p+q)^8$ (*Hint 3*). Expanded, we find that nine phenotypic classes are expected in the following ratio: 1 : 8 : 28 : 56 : 70 : 56 : 28 : 8 : 1. The F2 will be composed of the following tail lengths: 14, 12.5, 11, 9.5, 8, 6.5, 5, 3.5, 2 cm.

P18-3. P1: AABBCCDD (14 cm) aabbccdd (2 cm)

F1: AaBbCcDd (8 cm)

F2: All contributing alleles (14 cm), any seven contributing alleles (12.5 cm), any six contributing alleles (11 cm), any five contributing alleles (9.5 cm), any four contributing alleles (8 cm), any three contributing alleles (6.5 cm), any two contributing alleles (5 cm), any one contributing allele (3.5 cm), no contributing alleles (2 cm).

P18-4. Calculation of a correlation coefficient will provide an indication of the relationship between values for #1 and #2. If there is no relationship, the correlation will be 0. If there is a direct correspondence between #1 and #2, the correlation coefficient will be +1 or -1.

#1	#2	(#1-M1)	$(\#1-M2)^2$	(#2-M2)	$(\#2-M2)^2$	(#1-M1)(#2-M2)
8	5	1.33	1.769	-1.25	1.5625	-1.663
10	6	3.33	11.089	0.25	0.625	0.833
5	4	-1.67	2.789	-2.25	5.0625	3.758
8	10	1.33	1.769	3.75	14.063	4.988
3	5	-3.67	13.469	-1.25	1.563	4.586
8	7	1.33	1.769	0.75	0.563	0.998
4	7	-2.67	7.129	0.75	0.563	-2.003
9	4	2.33	5.429	-2.25	5.063	-5.243
6	4	-0.67	0.449	-2.25	5.063	1.508
5	7	-1.67	2.789	0.75	0.563	-1.253
6	7	-0.67	0.449	0.75	0.563	-1.253
8	9	1.33	1.769	2.75	7.563	3.658
Σ 80	Σ 75		Σ = 50.667		Σ = 42.813	Σ = 8.915

Mean 1 = 6.67 Mean 2 = 6.25

$s_1 = \sqrt{(50.667 / 11)} = 2.146$ $s_2 = \sqrt{(42.813 / 11)} = 1.973$

Cov (x,y) = 8.915 / 11 = 0.811

$r = 0.811 / (2.146)(1.973) = \mathbf{0.191}$

P18-5. Narrow-sense heritability can be calculated by dividing the observed correlation by the expected correlation. For full-siblings, the expected correlation is 0.5. $h^2 = 0.191 / 0.5 = 0.383$. Approximately 38% of the observed phenotypic variation in this population can be attributed to genetic factors. This estimate is specific to the population and the environment tested (*Hint 5*).

P18-6. Substitute the information provided into formula 18.8 (Tamarin) to obtain the increase in chocolate content expected: 0.69 = offspring gain / 12 mg

$$(0.69)(12 \text{ mg}) = \text{offspring gain} = 8.28 \text{ mg}$$

P18-7. If human height is controlled by the three pairs of additive alleles which add four inches to a base height of 4 feet, then there would be seven different phenotypes possible. An individual with no additive alleles would have the minimum height of 4 feet. An individual with any one contributing allele will be 4′ 4″. An individual with any two contributing alleles will be 4′ 8″ An individual with any three contributing alleles will be 5′. An individual with any four contributing alleles will be 5′ 4″. An individual with any five contributing alleles will be 5′ 8″. An individual with all six contributing alleles will be 6′. A woman with the minimum height would be genotype aabbcc while a male of intermediate would have genotype AaBbCc (or some other combination of three contributing and three noncontributing alleles). The woman will pass only noncontributing alleles to her offspring. Therefore the maximum height attainable for a child of this couple will be determined by the male parent. In this example, the maximum number of contributing alleles that can be passed to a child is three (ABC). Offspring of this couple cannot be taller than 5′.

P18-8. The expected correlations between relatives reflect the number of genes they share, on average. The correlations expected are: (a) 0.5; (b) 0.5; (c) 0; (d) 0.25; (e) 0.25; (f) 0.125; (g) 0.25

Answers to General Questions

Q18-1. Some potentially polygenic characteristics in human beings include height, weight, longevity, hair color, eye color, skin color, limb length, IQ, personality, aggression, memory, amino acid secretion, Alzheimer's disease, diabetes, fat : muscle ratio, blood pressure, lung capacity, heart rate, etc.

Q18-2. If there is no phenotypic variation in a population, then there is no measurable genetic variation ($V_P = V_G + V_E$; 0 = 0 + 0). Remember that heritability *measures the proportion of phenotypic variation* that is due to genetic factors. It does not determine to what extent a trait is genetic.

Answers to Even Numbered End of Chapter Exercises, Problems and Critical Thinking Questions

2. Two in five hundred is approximately 1 in 256 = $1/(4)^4$; therefore four loci (see table 18.1). With four loci and eight effective alleles maximum, each effective allele contributes about 1/4 inch above 2 inches. That is, the difference between 2 and 4 inches is 2 inches, divided by eight effective alleles is 1/4 inch per effective allele. For an approximate graph, see the right side of figure 18.1.

4. Polygenes are different from traditional Mendelian genes only in that several loci have a singular, additive effect on the phenotype. Through an additive model and environmental variability, a quantitative trait arises.

6. Begin by graphing the data. The general form of the data indicates that additive genes predominate. Excessive skew is caused by nonadditive effects. The number of loci involved is difficult to determine because the long-eared parent was so variable; more information is needed. (Probably four to eight loci are involved.)

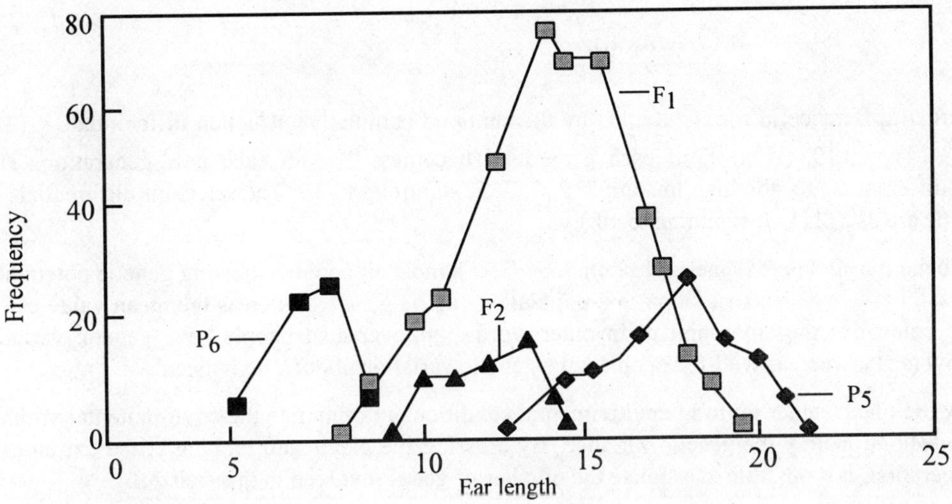

8. 3. If we assume that color is a quantitative trait, then the number of different phenotypes is given by $2n + 1$ where n is the number of loci. Since we see seven phenotypes, we probably have three different genes.

10. 6/16 or 3/8. This can be calculated from the binomial expansion. With two genes, the distribution among phenotypic classes is 1 (4 units):4 (3 units):6 (2 units):4 (1 unit):1 (0 unit).

12. The x variable is follicles from which we would like to predict the y variable, eggs laid. The x mean is 37.5; the y mean is 38.4.

$$\text{Cov}(x, y) = 113.2, s_x^2 = 163.0, s_y^2 = 116.6, r = 0.82, b\text{ (slope)} = 0.69, a\ (y\text{ - intercept}) = 12.3.$$

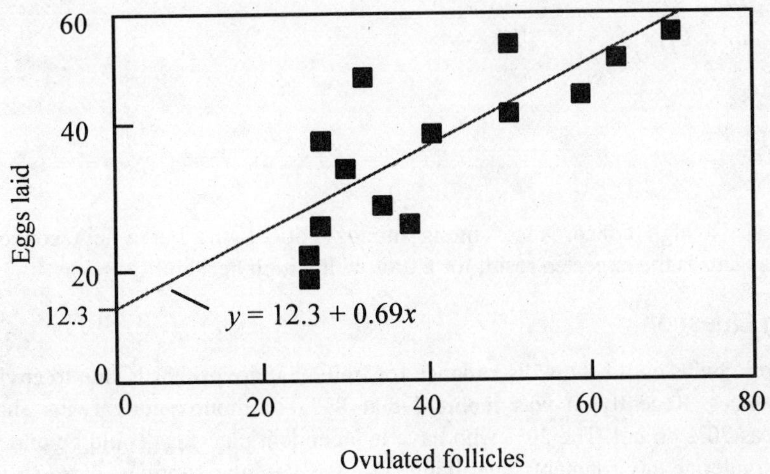

14. $H_B = 0.72$; $H_N = 0.45$. First calculate V_{ph} as 4.0 + 1.8 + 0.5 + 2.5 = 8.8. Broad sense heritability (H_B) is total genetic variance divided by total variance.

$$\frac{V_A + V_D + V_I}{V_{ph}} = \frac{4.0 + 1.8 + 0.5}{8.8} = \frac{6.3}{8.8} = 0.72$$

Narrow sense heritability (H_N) is additive genetic variance divided by total variance.

$$\frac{V_A}{V_{ph}} = \frac{4.0}{8.8} = 0.45$$

16. H = final difference in means divided by the summed cumulative selection differentials = (23 – 18)/(13 + 14) = 0.19. (Hint: Y_0 of each generation becomes $\bar{Y}$ with each new generation. Thus selection differentials for the high line are 3, 2, 3, 3, 2, summing to 13. The selection differentials for the low line are 3, 2, 2, 4, 3, summing to 14.)

18. Improved nutrition, medicine, and sanitation affect almost all people, allowing genetic potential to be universally affected. Thus variation in a population can be genetic, whereas the mean value can be under the control of the environment. In other words, impoverished people have genetic variation about a low mean, whereas well-fed people have genetic variation about a high mean.

20. Grow the plants under uniform environmental conditions to minimize the environmental variance in height, leaving mainly the genetic variance. By crossing the plants and looking at the extremes of the F_2 generation, it is possible to estimate the number of genes involved in this trait.

22. 0.7. This is calculated by using equation 18.8:

$$H = \frac{13.5 - 10}{15 - 10} = 3.5/5 = 0.7$$

24. $V_D = 35\ lb^2$; $V_E = 50\ lb^2$.

$H_N = V_A/V_{Ph}$

$V_A = 0.15 \times 100 = 15\ lb^2$

$V_G = V_A + V_D + V_I$

$50\ lb^2 = 15\ lb^2 + V_D = 0$

$V_D = 35\ lb^2$

$V_{Ph} = V_G + V_E$

$V_E = 50\ lb^2$

26. Yes. We see a high concordance among monozygotic twins but a low concordance among dizygotic twins. This is the expected result for a trait with a high heritability.

Critical Thinking Question

2. Monozygotic twins could show discordance for traits that are primarily due to environmental, not genetic, differences. Recently, it was reported that 80% of monozygotic twins share a common placenta whereas 20% do not. The 20% who have independent placentas could be under the influence of different intrauterine environments, involving different vascular supplies, more so than those who share a placenta. In addition, independent genetic changes take place during development such as in the maturation of B and T cells. Finally, developmental changes take place that do not cause or require genetics changes as in the development of neuronal connections in the brain.

CHAPTER 19

POPULATION GENETICS: THE HARDY-WEINBERG EQUILIBRIUM AND MATING SYSTEMS

Key Genetic Concepts

The Unit of Study: In population genetics, the focus changes from the products of individual matings to the characteristics of large interbreeding groups of individuals comprising a population. Populations are, as we will see in subsequent chapters, the units of evolutionary change. The complement of alleles available among all of the reproductive members of a population comprises the **gene pool**.

Allelic versus Genotypic Frequencies: To describe the genetic composition of a population, we can focus on genotypic frequencies (i.e. how many individuals are AA, Aa, or aa) or we can focus on the frequencies of specific alleles (i.e. what percentage of all alleles present are A). Allelic and genotypic frequencies are related: allele frequencies determine the genotypic frequencies if the population has an infinite number of members, random mating, no mutation, no migration, and no selection. When these conditions are met, the population is in **Hardy-Weinberg equilibrium**. The Hardy-Weinberg theory or rule states that allele frequencies of an autosomal gene will not change from one generation to the next and that the genotypic frequencies in the population can be predicted from allele frequencies. The Hardy-Weinberg law further states that, if a population is perturbed, equilibrium will be reestablished after one generation of random mating. The Hardy-Weinberg equation for one gene with two alleles is $p^2 + 2pq + q^2 = 1$.

Calculating Allele Frequencies: There are two general methods used to calculate allele frequencies. Which method is used depends on the nature of the data being examined. In all cases, the frequency of the dominant allele is denoted p and the frequency of the recessive allele is represented by q.

- In cases where heterozygotes can be identified, the direct count method is employed. In this case, allele frequencies are calculated by dividing the number of alleles of that type in the population by the total number of alleles in the population. For example, in a sample of 200 individuals there were 50 individuals with genotype AA, 100 with genotype Aa, and 50 individuals with genotype aa. The frequency of the A allele (p) can be calculated as: $(100 + 100) \div 400 = 0.5$, which corresponds to twice the number with the AA genotype + the number with heterozygous genotype ÷ twice the total number of individuals. Likewise, the frequency of the recessive allele (q) is 0.5. Notice that $p + q = 1$. Allele frequencies can be substituted into the Hardy-Weinberg equation to calculate genotypic frequencies expected if the population is in equilibrium. Chi-square analysis can be used to determine if the population is in equilibrium.
- In cases where the heterozygotes cannot be identified, Hardy-Weinberg equilibrium must be assumed to estimate allele frequencies. The frequency of the recessive genotype can be directly counted. This frequency corresponds to the q^2 portion of the Hardy-Weinberg equilibrium equation. The frequency of the recessive allele (q) can be obtained by taking the square root of the recessive phenotype frequency. The frequency of the

dominant allele (p) is obtained from the equation $p + q = 1$, by algebraic rearrangement ($p = 1 - q$).

Inbreeding: The assumption of random mating underlying Hardy-Weinberg equilibrium is violated when genetic relatedness (**inbreeding** or **common ancestory**) influences mating patterns. Inbreeding has several effects. First, it leads to an increased expression of hidden recessive alleles. Inbreeding increases the likelihood that a zygote will receive two copies of a deleterious allele. Second, inbreeding increases homozygosity. Inbred individuals are more likely to receive alleles that are copies of the same ancestral allele or **identical by descent** leading to **autozygosity**. A statistic referred to as the **inbreeding coefficient (F)** provides a measure of the probability of autozygosity. An F value of zero is obtained when there is no inbreeding while complete autozygosity (all alleles identical by descent) produces an F value of one. Pedigree analysis can be used to calculate an inbreeding coefficient for an individual. The inbreeding coefficient for a population can also be calculated by measuring the reduction in heterozygosity resulting from inbreeding.

Problem-Solving Hints

Hint 1: If heterozygotes can be identified, the direct count method should be used to calculate allele frequencies. *Never* assume equilibrium for a population in which heterozygotes can be identified.

Hint 2: Chi-square analysis can be used to determine if a population is in equilibrium. When using chi-square analysis, the degrees of freedom are obtained by subtracting *two* from the number of classes.

Hint 3: Hardy-Weinberg equilibrium must be assumed to calculate allele frequencies when heterozygotes cannot be identified. Obtain the frequency of the recessive allele (q) by taking the square-root of the frequency of the recessive phenotype.

Hint 4: At equilibrium, the distribution of genotypes AA, Aa, and aa is represented by p^2, $2pq$, and q^2, respectively.

Hint 5: The trinomial expansion can be used to calculate allele frequencies and expected genotypic frequencies for single genes with multiple alleles. For example, the Hardy-Weinberg equation for a gene with three alleles (p, q, and r) is: $p^2 + 2pq + 2pr + q^2 + 2qr + r^2 = 1$.

Hint 6: For a sex linked recessive trait, the frequency of affected males is equal to the frequency of the recessive allele (q) because males possess only one copy of the X chromosome (see Chapter 5). The frequency of affected females is q^2.

Hint 7: An excess of homozygotes in a population *suggests* inbreeding. Inbreeding alters genotypic but not allelic frequencies.

Practice Problems

Problem 19-1. Consider a population of cattle composed of 50 individuals with red hair (RR), 48 roan individuals (Rr), and 12 white individuals (rr). Calculate allele frequencies and determine if the population is in equilibrium.

Problem 19-2. In a human population, a recessive autosomal disorder occurs in approximately 1 of 2500 births. Calculate the frequency of the recessive allele and the percentage of heterozygous carriers.

Problem 19-3. In most states, newborn children are required by law to be tested for PKU, an autosomal recessive genetic disorder. If untreated, individuals homozygous for PKU suffer severe mental retardation. In a recent year, 4 phenylketonuric babies were detected out of 126,000 tested. Assuming H-W conditions, what is the frequency of the PKU gene in this population?

Problem 19-4. Consider the following data for allozyme phenotypes in the plant, *Phlox drummondii*: 7 AA, 15 AB, 38 BB, 30 BC, 16 CC, and 4 AC. Calculate the frequency of each allele and determine if the population is in equilibrium.

Problem 19-5. Assume that 30% of the human beings on campus have type A blood, 20% have type B blood, 10% have type AB blood, and 40% are type O. Calculate allele frequencies for this population.

Problem 19-6. Red-green colorblindness is a sex linked recessive trait in humans. Assume that the frequency of colorblind males on campus is 0.008. Calculate the frequency of all phenotypes.

Problem 19-7. Answer the above question assuming that the frequency given (0.008) pertains to colorblind females rather than males. What is the frequency of each phenotype in the population?

Problem 19-8. Assume that the frequency of heterozygotes in a population is 0.35. Assume further that $p = 0.65$ and $q = 0.35$. Estimate the inbreeding coefficient (F). Interpret the calculated value.

Problem 19-9. In forensic science, the product rule can be used to calculate the probability that a specific combination of alleles (usually from a crime scene sample) will be found in the an individual in the general population. As a result, a prosecutor can confidently tell a jury that the likelihood that someone other than the accused could have left the sample found at the crime scene is one in one billion, for example. Assume that a sample left at a crime scene has the genotype A1A2 B2B2 C1C2. The accused suspect also possesses the same genotype. Calculate the likelihood that someone other than the accused could have left the DNA sample found at the crime scene given the following allele frequencies: A1 = 0.04; B1 = 0.10; C1 = 0.20.

General Questions

Question 19-1. Since Hardy-Weinberg equilibrium assumptions are rarely met in nature, of what value are these assumptions.

Question 19-2: Tongue rolling is believed to be a dominant trait in human beings. A student of genetics was asked the following question: *Will the frequency of tongue rolling increase until 75% of the population expresses the dominant phenotype?* How would you answer this question?

Solutions to Practice Problems

P19-1. In this problem, heterozygotes can be identified and counted so we should use the direct method to calculate allele frequencies (*Hint 1*). The allele frequencies can be used to predict the expected genotypic frequency distribution at equilibrium.

total number of alleles = 2 (50 + 48 + 12) = 220
total number of r alleles = (12 x 2) + 48 = 72 $\quad$ f(r) = q = 72 / 220 = 0.327
total number of R alleles = (50 x 2) + 48 = 148 $\quad$ f(R) = p = 148 / 220 = 0.673

H_0 = population is in Hardy-Weinberg equilibrium

Genotype	Observed	Expected	$(O-E)^2$	$(O-E)^2 / E$
RR	50	p^2 x 110 = 49.82	0.0324	0.0007
Rr	48	2pq x 110 = 48.42	0.1764	0.004
rr	12	q^2 x 110 = 11.76	0.0576	0.005
				0.0097

Critical chi-square value (1 df, 0.05 probability; *Hint 2*) = 3.841 $\quad$ Fail to reject H_0

P19-2. In order to calculate allele frequencies, we must assume that the population is in Hardy-Weinberg equilibrium. The frequency of the recessive allele (q) is estimated from the frequency of the recessive phenotype. The percentage of heterozygous carriers is obtained by substituting the allele frequencies in the term 2pq.

f(rr) = 1/2500 = 0.0004 $\quad$ $q^2 = \sqrt{0.0004} = 0.02$ $\quad$ p= 1 - q = 0.98
f(Rr) = 2pq = 2(0.02)(0.98) = 0.0392 or 3.92%

P19-3. Approach this problem as you approached problem 19-2.

q^2 = 4/ 126,000 = 0.0000317 $\quad$ q = 0.006

P19-4. There are three alleles for this gene and we are able to identify all genotypic classes. Therefore, the direct count method is used to determine allele frequencies.

f(A) = p = [(7 x 2) + 15 + 4] / 220 = 0.15
f(B) = q = [(38 x 2) + 15 + 30] / 220 = 0.55
f(C) = r = [(16 x 2) + 30 + 4] / 220 = 0.3

H_0 = population is in Hardy-Weinberg equilibrium

Genotype	Observed	Expected	$(O-E)^2$	$(O-E)^2 / E$
AA	7	p^2 x 110 = 2.475	20.4756	8.273
AB	15	2pq x 110 = 18.15	9.9225	0.547
BB	38	q^2 x 110 = 33.275	22.3256	0.671
BC	30	2qr x 110 = 36.3	39.69	1.093
CC	16	r^2 x 110 = 9.9	37.21	3.759
AC	4	2pr x 110 = 9.9	34.81	3.516
				17.859

Critical chi-square value (4 df, 0.05 probability) = 9.488 $\quad$ Reject H_0

P19-5. In this problem, we must assume equilibrium to estimate allele frequencies (*Hint 3*). Obtain an estimate of r from the frequency of the recessive phenotype (type O blood). Next, examine the A and O, or B and O, alleles simultaneously to estimate a second allele frequency. The third frequency is obtained by subtracting from one.

$r = \sqrt{0.4} = 0.633$
$(p + r)^2 = p^2 + 2pr + r^2$
$(p + r)^2 = f(I^AI^A) + f(I^AI^O) + f(I^OI^O)$
$(p + r)^2 = 0.30 + 0.4$
$(p+r)^2 = 0.7$
$(p+r) = \sqrt{0.7} = 0.837$
$p = 0.837 - r$
$p = 0.837 - 0.633 = 0.204$
$q = 1 - 0.633 - 0.204 = 0.163$

P19-6. For a sex linked recessive trait, the frequency of affected males is equal to q (*Hint 6*). In this problem, q = 0.008. The frequency of unaffected males is 1 - 0.008 = 0.992. The frequency of colorblind females is q^2, or 0.0064%. The frequency of females with normal vision is $p^2 + 2pq$, or 99.99%.

P19-7. For a sex linked recessive trait, the frequency of affected females is equal to q^2. In this problem, $q^2 = 0.008$ so q = 0.089. The frequency of colorblind males is q, or 8.9% and the frequency of males with normal vision is 91.1%. The frequency of females with normal vision is $p^2 + 2pq$ or 99.2%.

P19-8. The inbreeding coefficient (F) is calculated by dividing the observed frequency of heterozygotes by the expected frequency of heterozygotes and subtracting this number from one.

$F = 1 - [0.35 / (2pq)]$
$F = 1 - [0.35 / 0.455]$
$F = 1 - 0.769$
$F = 0.231$

This value indicates that there is an excess of homozygotes that may be attributable to inbreeding (*Hint 7*).

Answers to General Questions

Q19-1. The Hardy-Weinberg equilibrium assumptions are valuable for several reasons. First, the concept provides us with a null hypothesis that can be tested. We cannot easily examine the forces of evolutionary change without knowing if the population is experiencing change or not. Second, assumptions of Hardy-Weinberg are important in human genetics because they allow the geneticist to calculate the frequency of a recessive allele in the general population. This method is the only one available for disorders that are caused by, as yet unidentified genes.

Q19-2. Many people mistakenly believe that 75% of the population will express a dominant phenotype given Mendelian ratios. Mendelian ratios are achieved in a population only when the frequency of each allele is 0.5. Substitute these numbers into the Hardy-Weinberg equation and you obtain Mendel's 1:2:1 genotypic and 3:1 phenotypic ratios. It is unreasonable to assume that alleles for all genes occur equally frequently.

Answers to Even Numbered End of Chapter Exercises, Problems and Critical Thinking Questions

2. a. We cannot count genes to determine allelic frequencies. If we assume Hardy-Weinberg equilibrium: $p = f(A)$, $q = f(a)$. Then there are nine *aa* individuals, or $f(aa) = 0.09$. If the Hardy-Weinberg equilibrium is assumed, $q^2 = f(aa) = 0.09$. Therefore $q = \sqrt{0.09} = 0.3$ and $p = 0.7$. Genotypic frequencies, assuming random mating, will be $f(AA) = p^2 = (0.7)^2 = 0.49$; $f(Aa) = 2pq = 2(0.7)(0.3) = 0.42$; and $f(aa) = q^2 = (0.3)^2 = 0.09$. Since there are zero degrees of freedom (number of phenotypes – number of alleles = 2 – 2 = 0), we cannot do a chi-square test.

b. The frequencies of the three genotypes are as follows: $f(FF) = 137/420 = 0.326$; $f(FS) = 196/420 = 0.467$; and $f(SS) = 87/420 = 0.207$. The allelic frequencies are $p = f(F) = f(FF) + (1/2)f(FS) = 0.326 + 0.467/2 = 0.560$; and $q = f(S) = 1 - 0.560 = 0.440$. We can thus set up the following chi-square table:

	FF	*FS*	*SS*	Total
Observed	137	196	87	420
Expected	$p^2 \times 420$	$2pq \times 420$	$q^2 \times 420$	
	131.5	207.0	81.5	420
Chi-square	0.230	0.585	0.371	1.186

The critical chi-square, at 0.05 with one degree of freedom, is 3.841. We thus fail to reject the null hypothesis that the population does not deviate from Hardy-Weinberg proportions.

4. Due to dominance, we must assume Hardy-Weinberg proportions. Assume *c* is the cystic fibrosis allele and *C* is the normal allele. Then $f(cc) = 1/2{,}500 = 0.0004$; $q = f(c) = \sqrt{0.0004} = 0.02$; and $p = f(C) = 1 - 0.02 = 0.98$. The frequency of heterozygotes is $2pq = 2(0.02)(0.98) = 0.0392$. Or 3.92% of people carry the cystic fibrosis allele.

6. a. In order to prove that the perpendicular to the base divides it into the ratio of *p* and *q*, we use the following diagram. In this diagram we need to prove $x/y = p/q$, or

$$\frac{x}{y} = \frac{\mathrm{D} + (1/2)\mathrm{H}}{\mathrm{R} + (1/2)\mathrm{H}}$$

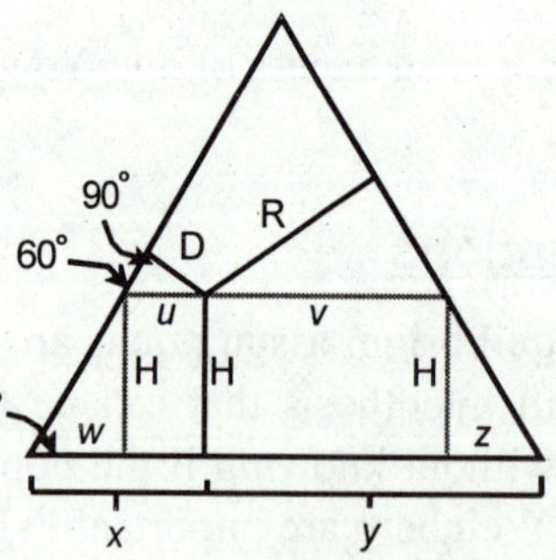

Remember that the three perpendiculars, D, R, and H, are the proportions of the three genotypes, *AA*, *aa*, and *Aa*, respectively. Thus D + (1/2)H = *p*, by definition as is *q* = R + (1/2)H. We do the following manipulation:

$$x = u + w$$

Since

$$\sin 60^\circ = \sqrt{3/2} = D/u, u = 2D/\sqrt{3}$$

$$\text{Tan } 60^\circ = \sqrt{3} = H/w, w = H/\sqrt{3}$$

Thus

$$x = u + w$$

$$x = 2D/\sqrt{3} + H/\sqrt{3}$$

or

$$x\sqrt{3}/2 = D + (1/2)H$$

Similarly

$$y = v + z$$

and

$$y\sqrt{3}/2 = R + (1/2)H$$

Dividing

$$\frac{(x\sqrt{3}/2)}{(y\sqrt{3}/2)} = \frac{D + (1/2)H}{R + (1/2)H}$$

or

$$\frac{x}{y} = \frac{D + (1/2)H}{R + (1/2)H}$$

b. To prove that the populations in Hardy-Weinberg equilibrium are a parabola:

At equilibrium: $D = p^2$; $R = q^2$; $H = 2pq$

Thus: $(2pq)^2 = 4p^2q^2$; or $H^2 = 4DR$; or $H^2 - 4DR = 0$

This is the general form of the equation of a parabola, $y^2 - 2ax = 0$.

8. $p = 0.2$, $q = 0.8$. We need to set up an equation that satisfies the condition: $q^2 = 2(2pq) = 4pq$. Substitute $1 - q$ for p:

$$q^2 = 4q(1 - q) = 4q - 4q^2$$

$$0 = 4q - 5q^2$$

$$0 = q(4 - 5q)$$

Either $q = 0$ or $5q = 4$; q cannot equal 0, or there would be no recessive homozygotes. So, q must = 0.8. Check by substituting: $q^2 = 0.64$, $2pq = 2(0.2)(0.8) = 0.32$.

10. $M = 0.43$; $N = 0.57$. No, it is not in Hardy-Weinberg proportions. If it were, we would expect $(0.57)^2$ *NN* individuals = 481.7, $(0.43)^2$ *MM* individuals = 281.7, and 2(0.43)(0.57) *MN* individuals = 736.7. The observed numbers are very different from the expected, and a chi-square analysis confirms this. (Chi-square, one degree of freedom = 527.049; $p << 0.01$.)

12. *BB* = 0.305; *Bb* = 0.578; *bb* = 0.117. For this problem you have to calculate allelic frequencies for males and females separately. Let $p = f(A)$, $q = f(a)$. Counting alleles: for the breeding women, $p = (20 + 6)/32 = 0.8125$ and $q = 6/32 = 0.1875$; for the males, $p = (4 + 8)/32 = 0.375$ and $q = 20/32 = 0.625$. If we assume random mating among males and females, we can set up a Punnett square to calculate offspring.

Female	*Male* 0.375 *B*	0.625 *b*
0.8125 *B*	0.305 *BB*	0.508 *Bb*
0.1875 *b*	0.070 *Bb*	0.117 *bb*

14. Let $p = f(A)$; $q = f(B)$; and $r = f(O)$. Due to dominance, we cannot count alleles and therefore must assume Hardy-Weinberg proportions. Thus $r = \sqrt{f(OO)} = \sqrt{0.09} = 0.3$. Type A + type O include the *AA*, *AO*, and *OO* genotypes. Therefore, assuming random mating, $(p + r)^2 = 0.07 + 0.09 = 0.16$; and $p + r = \sqrt{0.16} = 0.4$; $p = 0.4 - r = 0.4 - 0.3 = 0.1$. And, $q = 1 - p - r = 1 - 0.3 - 0.1 = 0.6$. However, since there is one degree of freedom (number of phenotypes – number of alleles = 4 – 3 = 1), we can do a chi-square test:

	Type A	*Type B*	*Type AB*	*Type O*	*Total*
Observed	7	72	12	9	100
Expected	$(p^2 + 2pr) \times 100$	$(q^2 + 2qr) \times 100$	$2pq \times 100$	$r^2 \times 100$	100
	$(0.01 + 0.06) \times 100$	$(0.36 + 0.36) \times 100$	0.12×100	0.09×100	
	7	72	12	9	

Since the observed and expected values are identical, the chi-square value is zero. The population thus appears to be mating at random.

16. First we determine allelic frequencies by counting alleles.

$$p_A = f(A) = (58 + 12)/100 = 0.7$$

$$q_A = f(a) = 1 - p_A = 0.3$$

$$p_B = f(B) = (58 + 22)/100 = 0.8$$

$$q_B = f(b) = 1 - p_B = 0.2$$

We calculate the expected linkage arrangement, assuming linkage equilibrium, by simple probability, to generate the following chi-square table:

	AB	*ab*	*Ab*	*aB*	Total
Observed	58	8	12	22	100
Expected	$p_A p_B \times 100$	$q_A q_B \times 100$	$p_A q_B \times 100$	$q_A p_B \times 100$	
	(0.7)(0.8)(100)	(0.3)(0.2)(100)	(0.7)(0.2)(100)	(0.3)(0.8)(100)	
	56	6	14	24	100
Chi-square	0.071	0.667	0.286	0.167	1.833

The critical chi-square (0.05, one degree of freedom) = 3.841. We thus fail to reject the null hypothesis of no deviation from linkage equilibrium.

18. $p = f(A) = 0.1$, $q = f(B) = 0.1$, $r = f(O) = 0.8$. Since we have three alleles, the distribution can be calculated as $(p + q + r)^2 = 1$. In this equation, we can let r^2 = frequency of type O individuals =

$$\frac{64}{100} = 0.64$$

Therefore, $r = 0.8$. Then $p^2 + 2pr + r^2 = (p + r)^2$ = frequency of type A + frequency of type O = 0.17 + 0.64 = 0.81;

$p + r = \sqrt{0.81} = 0.9$

Since $r = 0.8$, $p = 0.1$. Since $p + q + r = 1$, $q = 0.1$.

$$0.238.\ F = (2pq - H)/2pq = \frac{(0.42 - 0.32)}{0.42} = \frac{0.1}{0.42} = 0.238$$

20. There are three paths in this path diagram—one with B as the common ancestor (six ancestors in the path) and two with C as the common ancestor (one to D, with four ancestors, and one to E, with three ancestors). Although no other inbreeding coefficients are given, B is the offspring of half siblings (with A as the common ancestor) and has an inbreeding coefficient of $(1/2)^3 = 1/8$. Thus, summing the three paths:

$$F_1 = (1/2)^6(1 + 1/8) + (1/2)^4 + (1/2)^3 = 0.0176 + 0.0625 + 0.125 = 0.2051$$

22. There is one path, with A as a common ancestor. The path has a total of three ancestors. Given that F_A is 0.2, the inbreeding coefficient of G is $F_G = (1/2)^3(1.2) = 0.15$.

Path diagram

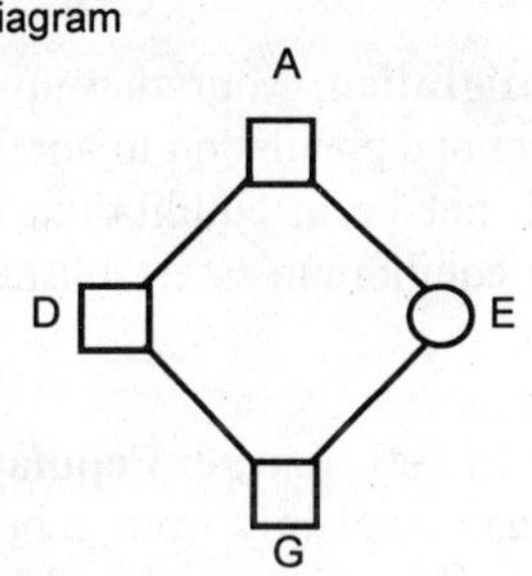

24. 0.238. First calculate allelic frequencies.

$$p = f(m) = \frac{2(14) + 32}{200} = 0.3$$

and $q = f(n) = 1 - f(m) = 0.7$. Inbreeding coefficient $F = (2pq - H)/2pq$, where H is observed frequency of heterozygotes. $2pq = 2(0.3)(0.7) = 0.42$; $H = 0.32$, so

$$F = \frac{(0.42 - 0.32)}{0.42} = 0.238$$

Heterozygotes have been reduced by about 24%, presumably because of inbreeding.

Critical Thinking Question

2. Many rare and endangered species are being bred in zoos to maintain their existence. However, the population sizes in zoos are inevitably small. Thus inbreeding due to small population size is inevitable. There are several results that follow. One is that the gene pool in the zoo population will not represent the gene pool in nature. Second, the inbreeding may seriously decrease the viability and vigor of the species. Thus the attempt is noble, but has potential drawbacks.

CHAPTER 20
POPULATION GENETICS: PROCESSES THAT CHANGE ALLELIC FREQUENCIES

Key Genetic Concepts

Stability of Equilibrium: Equilibria can be stable, neutral, or unstable. A stable equilibrium will return to the original point after it has been perturbed. A neutral equilibrium remains at the new allele frequency. An unstable equilibrium will not be reestablished after it is perturbed.

Violating the Assumption of No Mutation: Mutation changes one allele to another and thus is the source of new variation in populations. Mutation changes both allelic and genotypic frequencies over a period of *many* generations. Mutational equilibrium can be achieved and is stable.

Violating the Assumption of No Migration: Migration involves the adding or removing of alleles from populations as individuals move from one population to another. When migrants enter a native population, the resulting subdivided group will not be in equilibrium because of a decrease in heterozygotes (the **Wahlund effect**). Hardy-Weinberg equilibrium is established in the new conglomerate population after one generation of random mating.

Violating the Assumption of Infinitely Large Population: Small population size causes allelic frequencies in a population to fluctuate randomly from generation to generation. Random genetic drift is also referred to as "sampling error" because changes in allele frequencies result from the inexact sampling of the available alleles in a population. For example, a heterozygous individual may contribute the A allele to all of his/her offspring rather than to only 50% as expected. Genetic drift may be observed in a population founded by a small, nonrepresentative sample of a larger population. This phenomenon is referred to as a **founder effect**. The number of individuals in a population may be briefly reduced (a **bottleneck**). The small number of individuals that survive the bottleneck and reproduce afterwards may be nonrepresentative of the original population.

Violation of the Assumption of No Selection: Under Hardy-Weinberg equilibrium we assume that all members of a population are equally likely to reproduce. In reality, some phenotypes may leave more offspring than other phenotypes. This difference in **reproductive success** results from selection. Fitness is a measure of reproductive success: a genotype that leaves more offspring that survive and reproduce has a higher fitness than the alternative genotypes. Selection can act at different stages of the life cycle and can differentially affect survival (**zygotic selection**), gamete production (**gametic selection**), mating success (**sexual selection**), or fertility (**fecundity selection**). Selection acts upon the phenotype and in doing so changes the genotypic makeup of a population. In general, selection may shift the population mean toward one end of the phenotypic distribution (**directional selection**), it may favor individuals with phenotypes near the mean of the population (**stabilizing selection**), or it may favor individuals with extreme phenotypes (**disruptive selection**).

Selection is a necessary component of evolutionary change (Chapter 21). Selection may change allelic and genotypic frequencies by removing associated with a particular phenotype (for example a recessive phenotype). Once removed, the can be reintroduced into the population through mutation. Selection can also act to maintain two alleles in a population (**heterozygote advantage**). Consult your textbook for explanations of how to use genetic models to analyze the effects of selection.

Problem-Solving Hints

Hint 1: Models are useful for solving population genetic problems. First, set up the relevant algebraic model. Calculate q for the next generation. Calculate the change in "q" observed between generations. Calculate the value of q at equilibrium. Finally, determine if the equilibrium is stable.

Hint 2: Mutation, migration, genetic drift, and selection can change allele frequencies in populations.

Hint 3: Although mutation introduces new alleles into a population, its influence on allele frequencies is negligible if it acts alone.

Hint 4: Variation within populations is increased by mutation and migration while genetic drift decreases variation in small populations. Selection may increase or decrease genetic variation within a population.

Practice Problems

Problem 20-1. Assume that a normal allele mutates to a recessive allele at a rate of one in 10,000. Assume further that the recessive allele mutates back to the normal form at a rate of one in 200,000. The forward mutation rate is therefore twenty times the reverse mutation rate. If mutation acts alone, what allele frequencies are expected at equilibrium?

Problem 20-2. Consider a gene which shows overdominance: the heterozygote has a higher fitness in the environment than the homozygous normal individual (AA) and the homozygous affected individual (aa) has reduced fitness. Given the fitness values below, calculate allele frequencies.

Genotype	Fitness
AA	0.75
Aa	1.0
aa	0.1

Problem 20-3. Assume that school spirit is genetically controlled by a single gene with two alleles. Assume further that support for your school is determined by the dominant allele (S) while the recessive allele (s) in homozygous form results in support for a rival school. The frequency of the dominant allele is 0.9 at your school while the frequency of the dominant allele is 0.2 at the rival institution. In an effort to secure increased funding for your school, you and your classmates propose a scheme for increasing support for your institution: you plan to send migrants from your institution to the rival institution. You consider three different migration rates: 0.1, 0.5, and 0.8. Calculate the frequency of the S allele in the rival population after each migration event. Will each of the above migration rates alter the allele frequencies in the rival institution?

Problem 20-4. Assume that a recessive phenotype is selected against and that the selection coefficient is 0.1. The mutation rate (A to a) is 4×10^{-6}. What is the selection-mutation equilibrium frequency of the recessive allele?

Problem 20-5. For each major type of selection (directional, stabilizing, and disruptive), provide the relative fitness values associated with each genotype (AA, Aa, and aa).

General Questions

Question 20-1. For each force discussed in this chapter (genetic drift, migration, mutation, directional selection, stabilizing selection, and disruptive selection), speculate whether each would act to increase or decrease variation within populations and between populations.

Question 20-2. Would selection against a dominant allele or against a recessive allele be most effective in removing the allele from the population?

Solutions to Practice Problems

P20-1. Substitute the forward and reverse mutation rates into the equation:

q (equilibrium) = $\mu / (\mu + \nu)$ and solve for the allele frequency.

$$q \text{ (equilibrium)} = 20\nu / (20\nu + \nu) = 20\nu / 21\nu = 0.952$$

The frequency of the normal allele is obtained by subtracting the equilibrium value of q from one. In this case, p = 1 - 0.952 or 0.048.

P20-2. First, calculate the selection coefficients for the homozygous normal and the homozygous affected genotypes. Remember that the selection coefficient is equal to one minus fitness. In this problem, the selection coefficient for the homozygous normal genotype (s_1) is 1 - 0.75 = 0.25. The selection coefficient for the homozygous affected genotype (s_2) is 1 - 0.1 = 0.9. Next, calculate the frequency of q expected at equilibrium (= $s_1 / (s_1 + s_2)$). Substituting, we find that the frequency of q expected at equilibrium = 0.25 / (0.9 + 0.25) or 0.217. The frequency of p would therefore be 1- 0.217 or 0.783.

P20-3. The equation used to calculate the allele frequency in the rival population after the migration is: $p_c = p_n + m(p_m - p_n)$ where p_c = the new frequency in the population, p_n and p_m are the original frequencies in the rival and migrant populations respectively and m is the migration rate. As you can see below, each of the migration rates altered the allele frequencies in the rival population although large changes require movement of a high percentage of the students from your institution.

m = 0.1   0.2 + 0.1 (0.9 - 0.2) = 0.27
m = 0.5   0.2 + 0.5 (0.9 - 0.2) = 0.55
m = 0.8   0.2 + 0.8 (0.9 - 0.2) = 0.76

P20-4. The frequency of q at equilibrium equals: $\sqrt{(\mu/s)} = \sqrt{(4 \times 10^{-6} / 0.1)} = 0.006$. The frequency of the selected against allele (a) is very low, but it is maintained in the population by mutation.

P20-5. Directional selection can be modeled as selection against a recessive homozygote. Stabilizing selection can be modeled as heterozygote advantage while disruptive selection is modeled as heterozygote disadvantage. The table of fitness values for each type of selection are presented below.

Selection Type	AA	Aa	aa
Directional	1	1	1-s
Stabilizing	$1-s_1$	1	$1-s_2$
Disruptive	1	1-s	1

Answers to General Questions

Q20-1. Within a population, you would expect mutation, migration, and stabilizing selection to increase variation. Mutation and migration introduce new or rare alleles into a population while stabilizing selection maintains both alleles in a population. Genetic drift, directional selection, and disruptive selection act to remove alleles (directly or accidentally) and would therefore decrease variation within a population. Between populations, genetic drift, disruptive selection, and directional selection would act to increase differences. Mutation, migration, and stabilizing selection would tend to increase the similarity between different populations.

Q20-2. Selection against a dominant phenotype would quickly remove the dominant allele from the population. In contrast, selection against a recessive allele occurs *very* slowly. This is because heterozygotes carry a "hidden" recessive allele. Remember that selection acts on phenotypes . . . heterozygotes express the dominant phenotype and would not be selected against. Heterozygotes would live and be able to pass the recessive allele to their offspring. As the recessive allele becomes rare, fewer individuals will be homozygous recessive and selected against. With each generation it become harder to remove the recessive allele from the population.

Answers to Even Numbered End of Chapter Exercises, Problems and Critical Thinking Questions

2. When there is no back mutation, we have a simple model:

$$A \xrightarrow{\mu} a$$

Therefore

$$q_{n+1} = q_n + \mu p_n$$

$$\Delta q = q_{n+1} - q_n = q_n + \mu p_n - q_n = \mu p_n$$

At equilibrium:

$$\Delta q = 0 : \hat{p} = 0 \text{ and } \hat{q} = 1.0$$

4. 0.83. If μ and ν represent forward and reverse mutation rates respectively, the equilibrium frequency is given by $q = \mu/(\mu + \nu)$. In this case $\mu = 5\nu$, so $q = 5\nu/(5\nu + \nu) = 5/6 = 0.833$.

6. The population had a deficiency of heterozygotes. If natural selection is not acting on the population, then the deficiency could have resulted from inbreeding or subdivision—the Wahlund effect—or both. Further information would be needed to understand fully what is happening in this population.

8. About 47%. To calculate the migration frequency, use equation 20.12:

$$m = \frac{q_c - q_1}{q_2 - q_1} = \frac{(0.32 - 0.25)}{(0.4 - 0.25)} = \frac{0.07}{0.15} = 0.47$$

10. The ultimate fate of the *A* allele is either loss or fixation. The probability of loss is 0.3; one hundred generations; two thousand generations.

12.

	AA	*Aa*	*aa*	Total
Before selection	p^2	$2pq$	q^2	1
Fitnesses (*W*)	1	$1\text{-}s$	$1\text{-}1.5s$	
Ratio after selection	p^2	$2pq(1\text{-}s)$	$q^2(1\text{-}1.5s)$	$\overline{W} = 1\text{-}2pqs\text{-}1.5sq^2$
Frequencies after selection	$p^2/\overline{W}$	$2pq(1\text{-}s)/\overline{W}$	$q^2(1\text{-}1.5s)/\overline{W}$	1

Fitnesses could be 1, 1/2, and 1/4 for *AA*, *Aa*, and *aa*, respectively (or 1, $1 - s$, and $1 - 1.5s$, where $s = 0.5$):

Then

$$q_{n+1} = pq(1-s)/\overline{W} + q^2(1-1.5s)/\overline{W}$$

$$\Delta q = q_{n+1} - q$$

$$= pq(1-s)/\overline{W} + q^2(1-1.5s)/\overline{W} - q\overline{W}/\overline{W}$$

After some algebraic manipulation, this equation simplifies to

$$\Delta q = sq(q[1.5-0.5q]-1)/\overline{W}$$

At equilibrium (Δq = 0), and therefore the numerator has to equal zero. This is satisfied if $\hat{q}$ = 0 because q is a multiplier of the numerator. If $\hat{q}$ = 1, the expression in parentheses is zero; use trial and error or determine the genetically meaningful roots of the quadratic with:

$$\frac{-b \pm \sqrt{b^2 - 4ac}}{2a}$$

If we substitute various values of q into the Δq equation, we see that the equilibrium is stable. At any value of q other than $\hat{q}$ (zero or one), Δq will be negative, indicating a decrease in q in the next generation. Thus $\hat{q}$ = 1 is a trivial equilibrium point. Any population of individuals segregating both alleles will eventually lose the a allele.

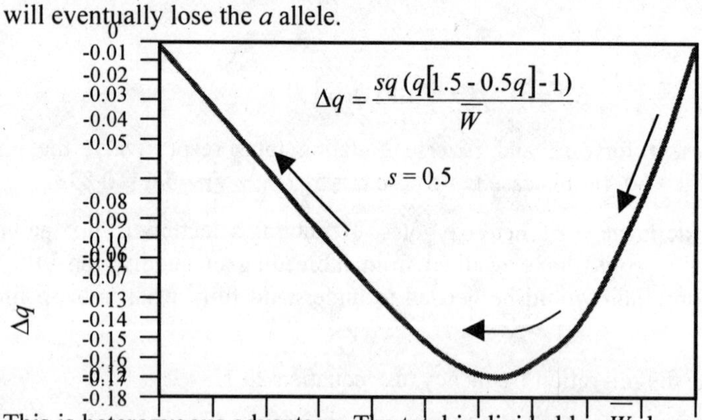

14. a. This is heterozygous advantage. The total is divided by $\overline{W}$ because losses have occurred in the homozygous categories.

b. $q_{n+1} = (pq + q^2[1 - s_2])/\overline{W}$

c. Assume $s_2 = 2s_1$. Then substituting into equation 20.24:

$$\hat{q} = s_1/(s_1 + s_2) = s_1/(s_1 + 2s_1) = s_1/3s_1 = 0.33.$$

Therefore,

$$\hat{p} = 0.67$$

If $s_1 = 0.1$ and $s_2 = 0.3$, then

$$\hat{q} = 0.1/(0.1 + 0.3) = 0.1/0.4 = 0.25; \; \hat{p} = 0.75$$

$$\text{Heterozygotes} = 2\,\hat{p}\,\hat{q} = 0.375 = 37.5\%$$

16. $\hat{q} \approx \sqrt{\mu/s}$; see equation 20.22

$\hat{p} \approx 1 - \sqrt{\mu/s}$

Substituting

$$\hat{q} = 0.018;\ \hat{p} = 0.982$$

18. By inspection, you can see that there has been a relative decrease in sepia-eyed flies and a relative increase in wild-type homozygotes and heterozygotes. We need to convert these changes to fitnesses. We do so as follows:

Fitness of s^+s^+ = 30/25 = 1.2

Fitness of s^+s = 60/50 = 1.2

Fitness of *ss* = 10/25 = 0.4

These "fitness" values can be changed so that the highest fitness is 1.0 by dividing each by 1.2, yielding fitnesses of 1, 1, and 0.33. This is then an example of selection against the homozygous recessive genotype (see table 20.2) in which $s = 0.67\ (1 - 0.33)$.

20. $q = 0.4375$; $p = 0.5625$. We set up the following selection model:

	AA	*Aa*	*aa*	Total
Initial	0.25	0.50	0.25	1.0
W	1	0.8	0.6	$\overline{W}$ = 1(0.25) + 0.8(0.5) + 0.6(0.25) = 0.8
Ratio after selection	1(0.25)	0.8(0.5)	0.6(0.25)	
	0.25	0.40	0.15	0.8
Genotypic frequencies after selection				
	$0.25/\overline{W}$	$0.4/\overline{W}$	$0.15/\overline{W}$	1.0
	0.25/0.8	0.4/0.8	0.15/0.8	1.0
	0.3125	0.50	0.1875	1.0

$$q_{n+1} = f(aa) + (1/2)f(Aa) = 0.1875 + 0.25 + 0.4375$$

$$p_{n+1} = 1 - q_{n+1} = 1 - 0.4375 = 0.5625.$$

22. 0.0158. We use equation 20.22:

$$\hat{q} = \sqrt{u/s} = \sqrt{\frac{5 \times 10^{-5}}{0.2}} = 0.0158$$

Critical Thinking Question

2. Whenever heterozygote deficiency is observed, scientists immediately think of inbreeding. When that is eliminated as a cause, the Wahlund effect is the next obvious explanation. Finally, selection could be eliminating heterozygotes. A closer study of the species under question is necessary to separate these possibilities.

CHAPTER 21
EVOLUTION AND SPECIATION

Key Genetic Concepts

Darwinian Evolution: Evolution describes the changes in allele frequencies in populations over generations. While Charles Darwin did not discover evolution, he did propose a mechanism of evolution, **natural selection**. The requirements for evolution by natural selection include variation among individuals, inheritance of these differences, and differential reproduction of individuals possessing these traits.

Speciation: Speciation refers to a process whereby an ancestral species evolves into a new species (**anagenesis**) or whereby one population diverges to form two, simultaneously existing species (**cladogenesis**). In order for speciation to occur, populations must be isolated and members must be unable to reproduce with each other (**reproductive isolation**). Environmental, behavioral, mechanical, or physiological barriers prevent isolated populations of individuals from interbreeding and exchanging genes (gene flow). Over evolutionary time, the separated members of one species become two distinct species incapable of reproducing with each other. A number of prezygotic (before zygote formation) mechanisms have been identified (see textbook for a listing). Speciation can occur while populations are separated by geographic barriers (**allopatric speciation**), spatial and/or temporal barriers (**parapatric speciation**), or by polymorphisms within the population (**sympatric speciation**).

Gradualism versus Punctuated Equilibrium: Evolution, as envisioned by Darwin, is a gradual process (**phyletic gradualism**). Examination of the fossil record suggests that evolution occurs in rapid bursts separated by long periods of stasis (**punctuated equilibrium**).

Variation: When more than one allele occurs at a given locus (at a frequency of 5% or more), we describe the locus as **polymorphic**. Many loci in a variety of organisms are polymorphic. Variation within populations can be maintained through selection against both homozygotes (**heterozygote advantage**), through **frequency-dependent selection** in which genotype fitnesses change according to their frequencies in the population, or because the polymorphic loci is neutral or not under selective pressure. **The neutral gene hypothesis** has been proposed as an alternative to the natural selection model. Neutralists argue that variation in a population results from the accumulation of mutations and genetic drift.

Molecular Evolutionary Clocks: A molecular evolutionary clock measures evolutionary time in the number of nucleotide substitutions per million years. Statistical methods can be used to estimate the number of nucleotide substitutions that have taken place over evolutionary time if the current amino acid differences in a protein between species is known.

Sociobiology: Sociobiology is an emerging field that is based on the premise that social behavior is under genetic control. Specifically, sociobiology focuses on the evolution of altruistic behavior: behavior that lowers the fitness of the performer while increasing the fitness of the recipient. **Altruism** is compatible with individual selection if one focuses on the total genetic contribution of an individual to the next generation. Obviously, an organism passes its genes to the next generation directly through the production of offspring (fitness). Not so obvious is the indirect passing of an individual's genes through the reproduction of relatives. **Inclusive fitness** refers to the sum of an individual's fitness plus the fitness gained through helping relatives produce offspring. If we consider inclusive fitness, we see that sacrificing reproduction to help a relative reproduce can be favored by natural selection if the end result is the individual contributes a greater number of copies of their alleles to the next generation. This type of selection is referred to as **kin selection**. Inclusive fitness and kin selection can even explain the existence of sterile castes in the truly social bees, ants, and wasps.

Problem-Solving Hints

Hint 1: Mutation is responsible for introducing alleles into a population. Alone, mutation acts slowly alone although an allele can spread rapidly if it confers a selective advantage.

Hint 2: Individual selection is a "stronger" force than group selection. Most cases of traits that appear to be "for the good of the species" can actually be best explained in terms of individual selection.

Hint 3: Altruism may be favored by natural selection if the individual that sacrifices reproduction increases their genetic contribution to the next generation as a result of the altruistic act. Inclusive fitness applies only to the number of offspring directly resulting from the help provided to a genetic relative.

Hint 4: When inferring evolutionary relationships, the fewest number of chromosomal changes should be considered.

Practice Problems

Problem 21-1. Identify each of the reproductive isolating mechanism described below.

a. Two species of insect produce different chemicals used to locate mates.

b. Two species of insect produce the same mating signal but one mates at dusk and the other mates at dawn.

c. Two species produce phenotypically normal offspring that are sterile.

d. The pedipalp if a male spider (a sperm transfer structure) is shaped differently than the female reproductive tract.

e. Two species produce compatible gametes but one species lives in the forest canopy while the other species occurs on the forest floor.

Problem 21-2. T.H. Huxley was reportedly asked if he would risk his life to save the life of others. His reported answer? Yes, but only to save two of his brothers, four of his uncles, and eight of his cousins. Explain.

Problem 21-3. On an expedition to Madagascar, you find three isolated, phenotypically identical populations of zeebogs. Upon cytological examination, you find that each population differs with regard to the sequence of genes found on chromosome number 10. Specifically, the genes Z1, Z2, Z3, Z4, Z5, and Z6 are all present, although their sequence differs between populations. Which population is likely ancestral to the others? Note, the centromere is represented by the letter "c."

Population 1:	Z1	Z2	c	Z4	Z3	Z5	Z6
Population 2:	Z1	Z3	Z4	c	Z2	Z5	Z6
Population 3:	Z1	Z2	c	Z4	Z6	Z5	Z3

General Questions

Question 21-1. Punctuated equilibrium is often referred to as "evolution by jerks." Why is this an accurate description of the process?

Question 21-2. The three requirements for evolution by natural selection are variation, selection, and inheritance. Elaborate on the necessity of each factor.

Question 21-3. A friend once posed (although not seriously) that her relationship with her step-children was strained because of an underlying, subconscious concern about her fitness. Explain.

Solutions to Practice Problems

P21-1. To identify which reproductive isolating mechanism is being described you must determine how the two populations are being separated. For example, the two species described in (a) use different types of mating signals which suggests behavioral isolation. (b) Different times of mating suggest temporal isolation. (c) Production of sterile offspring results from hybrid sterility. (d) Incompatibility of reproductive structures is a type of mechanical isolating mechanism. (e) Physical separation of two compatible species is an example of geographic isolation.

P21-2. This problem asks you to consider the concept of inclusive fitness. A sacrificing individual could be favored if his/her sacrifice leads to an equal or elevated genetic contribution to the next generation (*Hint 3*). Again, relatives share alleles in common. Therefore, indirect fitness gains may be possible. The correlation between siblings is 0.5. In other words, siblings share, on average, 50% of their genes. Two siblings would "equal" Huxley's loss of his life. The same applies for uncles which share approximately 25% of their genes with their nephew and for cousins which share 12.5% of their genes with Huxley.

P21-3. To infer the evolutionary relationships between the three populations, you need to determine how many changes would be necessary if each of the species were ancestral. The order requiring the smallest number of changes is the most likely (*Hint 4*). Population 2 is probably ancestral. One inversion (of the Z3 Z4 Z2 sequence) would produce population 1. An inversion in population 2 (of the Z3 Z5 Z6 sequence) would produce population 3.

Answers to General Questions

Q21-1. Punctuated equilibrium is not referred to as "evolution by jerks" because the phrase describes the people who believe it! The "jerks" mentioned refer to the rapid periods of evolution which are followed by long periods of stasis in this model.

Q21-2. Natural selection can act only on phenotypic variation present in a population (*Hint 1*). If variation is not present, there is no selective advantage to any one phenotype. Natural selection also requires that the source of the variation selected for or against have a genetic component. Allele frequencies in a population cannot change over generations if there is no mechanism for transmitting the observed differences from parents to offspring. Natural selection is the mechanism of change. Without selection, all phenotypes would be equally likely to reproduce and leave offspring.

Q21-3. Although this is an unusual question, it is relevant to the discussion of fitness and inclusive fitness (*Hint 3*). An organism that invest time rearing young that are genetically unrelated reduces its own fitness. Investments in such individuals reduces the investment that can be made in the individual's genetic offspring. Inclusive fitness is gained only if the individual benefiting from the sacrifice is genetically related to the helper. In humans, step children are rarely genetically related to the step parent. As you read in the textbook chapter, applying the concepts of sociobiology to humans is controversial. I think that this question illustrates why this is the case.

Answers to Even Numbered End of Chapter Exercises, Problems, and Critical Thinking Questions

2. By definition, a population increase is a period during which more young are surviving per adult than normal. Whatever the circumstances are (e.g., new food supply, favorable weather, reduction in predators), competition is relaxed and therefore offspring that normally would not survive (under stable numbers) do survive. This does not mean that all selection is relaxed. For example, fecundity selection can be intense.

4. Evolutionary rates determined from the protein, mtDNA, or nuclear DNA clocks are calculated assuming average rates of change. Looking at the end product (contemporary species) can provide insights into the actual rate at which species diverged by comparing rates in different lineages of known time of divergence. However, fluctuations in rate may be impossible to determine from extant species. In order to settle this controversy, much more detailed fossil lineages are needed.

6. Because cows and human beings are different species, isolating mechanisms almost certainly exist. In this situation, we expect mechanical isolation—failure of bull sperm to reach the human egg or to fertilize it, and hybrid weakness—the hybrids probably fail to form or abort because of different developmental signals—to be operating.

8. The hybrids will have twenty-one chromosomes (three sets of seven). Three chromosomes cannot align and segregate equally, so various types of aneuploids will be produced in both gametes and zygotes. These aneuploids usually will not survive.

10. In this example, $p = f(F) = 0.7$ and $q = f(S) = 0.3$. The expected distribution is *FF*, forty-nine; *FS*, forty-two; and *SS*, nine. The chi-square value is 8.163, one degree of freedom, indicating that this population is not in Hardy-Weinberg proportions. There are too many heterozygotes and too few homozygotes, indicating heterozygote advantage. Frequency-dependent selection cannot be determined from one sample; we would have to see a change in selection coefficients as allelic frequencies changed. However, frequency-dependent selection would not look like heterozygote advantage. Rather, it would look like directional selection favoring the rare allele. A transient polymorphism is also an example of directional selection and would not look like heterozygote advantage.

12. In the neutral gene hypothesis, it is suggested that polymorphisms occur in populations due to a balance of mutation and random genetic drift of neutral or nearly neutral alleles. The alternative view is that all variation is maintained by selective mechanisms. To distinguish these views, it is necessary to either look for selection in a large number of loci or determine some global pattern in distribution of allelic frequencies. Clearly both views are correct some of the time.

14. The investigators have suggested that this is a form of Batesian mimicry in which the normal leaves are mimicking those that have been attacked by caterpillars. In this way other butterflies and moths (and other insects) may be deterred from laying their eggs on these leaves to avoid the competition from larvae (caterpillars) thought to be there already.

16. Natural selection leads to adapted organisms; neutralism does not necessarily change that view. Neutralism is a theory to account for the tremendous amount of molecular variation that occurs in natural populations, variation that until the 1960s was not known to be present.

18. 5×10^7 years. In this problem, $K = 0.2$ and $k = 2 \times 10^{-9}$. Thus, $k = K/2T$, so $T = K/2k = 0.2/2(2 \times 10^{-9}) = 0.5 \times 10^7$ years.

20. No. Certain proteins, and hence certain coding sequences, are likely to be conserved. Since most introns seem to have no function, more mutations in introns would be expected to accumulate over time. Since cDNA does not contain introns, we underestimate the number of base changes, and species seem to be more closely related than they may be.

22. Individual selection is defined by the fitnesses of genotypes of individuals, along the lines of the selection models that we looked at in chapter 20. Group selection is a concept, true under only a few circumstances, in which natural selection favors actions by individuals that are beneficial to the group at the expense of the fitness of the individuals. Kin selection is natural selection favoring behavior in individuals that benefits the individuals and their relatives that carry similar alleles. Altruistic acts are explained most easily under individual selection as selfish acts. Group selection, which rarely happens in nature, explains altruism as the favoring of groups that exhibit it. Kin selection explains altruism by an increase in inclusive fitness.

24. Individuals are driven to preserve and enhance the frequency of their own alleles in the population. We expect, and this has been confirmed in many natural populations, that males will kill the young of other males but not those they fathered.

Critical Thinking Question

2. Wynne-Edwards explanation makes much intuitive sense but is simply wrong because nature does not work that way. Any bird that lived under that system but had a gene that caused it to breed even when densities are high would produce more young than the birds that ceased reproduction for the good of the species. Thus in the next generation, there would be more of the "selfish" gene and less of the "altruistic" alternative. Thus altruism is selected against. The birds might cease reproducing at high densities, but they are doing it for selfish reasons. Probably success in rearing young is very low when resources are low and a bird that waits will have more energy for successful breeding next time.